Taking Sudoku Seriously

Taking Sudoku Seriously

The Math Behind the World's Most Popular Pencil Puzzle

JASON ROSENHOUSE

AND

LAURA TAALMAN

OXFORD
UNIVERSITY PRESS

Oxford University Press, Inc., publishes works that further
Oxford University's objective of excellence
in research, scholarship, and education.

Oxford New York
Auckland Cape Town Dar es Salaam Hong Kong Karachi
Kuala Lumpur Madrid Melbourne Mexico City Nairobi
New Delhi Shanghai Taipei Toronto

With offices in
Argentina Austria Brazil Chile Czech Republic France Greece
Guatemala Hungary Italy Japan Poland Portugal Singapore
South Korea Switzerland Thailand Turkey Ukraine Vietnam

Published by Oxford University Press, Inc.
198 Madison Avenue, New York, New York 10016
www.oup.com

Library of Congress Cataloging-in-Publication Data

Rosenhouse, Jason.
Taking sudoku seriously : the math behind the world's most popular pencil puzzle /
Jason Rosenhouse and Laura Taalman.
p. cm.
Includes bibliographical references.
ISBN 978-0-19-975656-8
1. Sudoku. 2. Mathematics—Social aspects. I. Taalman, Laura. II. Title
GV1507.S83R67 2012
793.74—dc22 2011003188

9 8 7 6 5 4 3 2 1

Printed in China on acid-free paper

In memory of Martin Gardner, who showed a generation of mathematicians the value of puzzles as a gateway into mathematics.

CONTENTS

PREFACE

Every math teacher knows the frustration of directing a seemingly simple question to a class and receiving blank stares in return. In part, this reaction can be attributed to general student apathy or to a fear of giving the wrong answer. There is, however, a more fundamental issue to be addressed.

Most people, when asked to describe mathematics, will talk about the tedious algorithms of arithmetic or the seemingly arbitrary rules of algebra. For them it is all about symbol manipulation and mindless computation. This view is entirely understandable given that they probably saw little else in their grade school and high school mathematics classes.

Mathematicians do not recognize their discipline in such descriptions. We see arithmetic and algebra as tools used in doing mathematics, just as hammers and handsaws are tools used in carpentry. For professionals, mathematics is about curiosity, imagination, and solving problems. There are questions that are instinctive and natural for mathematicians that rarely occur to those looking in from the outside. There is such a thing as a mathematical view of the world. Sadly, it is a view that is too often hidden from those struggling to learn the subject.

Which brings us back to the blank stares. Often the problem is simply that mathematicians have a way of expressing themselves that makes little sense to those outside the club. Students unaccustomed to the sorts of questions mathematicians ask, or unaware that mathematics is about asking questions in the first place, will often be confused by questions more experienced people regard as simple. We need first to develop mathematical thinking in our students before we expect them to toss off answers to our questions.

That is where the Sudoku puzzles come in. We can define a *Sudoku square* as a 9×9 grid in which every row, column, and 3×3 block contains the digits 1–9 exactly once. A *Sudoku puzzle* is then a square in which some of the cells have been filled in while others are blank. The goal of the solver is to fill in the blank cells in such a way that the result is a Sudoku square. If the puzzle is sound there will only be one way of doing that.

Here is an example. This is a level 3 puzzle, where level 1 is the easiest and level 5 is the hardest.

Puzzle 1: Sudoku Warm-Up.

Fill in the grid so that each row, column, and block contains each of the numbers 1–9 exactly once. The solution to this puzzle is at the end of the book.

	2	3						
	5		7				8	3
		4				2		7
				7			6	
			9		6			
	3			2				
3		5				8		
6	1				8		7	
						4	1	

Over the past five years, Sudoku puzzles have become a mainstay of many newspapers. Such venues are typically careful to assure the reader that, the presence of numbers notwithstanding, Sudoku puzzles are not math problems. They are keen to stress that any collection of nine distinct symbols, such as the first nine letters of the alphabet, would work just as well.

This sort of thing sounds bizarre to a mathematician. In saying that Sudoku does not involve mathematics, the newspaper really means it does not involve arithmetic. The sort of reasoning that goes in to solving a Sudoku puzzle, on the other hand, is at the heart of what mathematics is all about. That so many people will claim to hate doing mathematics while simultaneously enjoying the challenge of solving a puzzle is a source of frustration to those of us in the business.

To a mathematician, Sudoku puzzles immediately suggest a whole host of interesting questions even beyond the reasoning that goes in to solving them. How many Sudoku squares are there? What sorts of transformations can you do to a Sudoku square to produce other such squares? What is the smallest number of initial clues a sound puzzle can have? What is the largest number of initial clues a puzzle can have without having a unique solution? Is it possible to have a Sudoku square in which each 3×3 block is actually a semimagic square (so that the digits in each row and column within the block add up to the same sum)? In attempting to answer these questions we will inevitably encounter a lot of interesting mathematics.

More than that, however, we will use Sudoku puzzles and their variants as a gateway into mathematical thinking generally. This is both a math book and a puzzle book. The puzzles, in addition to being enjoyable simply as stand-alone brainteasers, will serve to complement and introduce the mathematical concepts in the text. Our emphasis throughout is on asking questions and solving problems; technical mathematical machinery will be introduced only as it arises naturally in the course of our reasoning.

We have a number of different audiences in mind. For students in high school or college we intend to provide a view of mathematics that is very different from what is usually presented. It is a far more realistic view than the one implied by years of training in tedious symbol manipulation. For educators we hope to provide some novel ideas for how to bring genuine mathematical thinking into the classroom in a context that will be interesting and accessible to students. For any layperson with a general interest in mathematics, we provide plenty of food for thought and intellectual stimulation. Professional mathematicians can benefit from seeing familiar mathematical abstractions applied in novel settings.

We have assumed little beyond high school mathematics. Indeed, if you flip through the book right now you will notice that for the most part we make limited use of mathematical symbols. Our focus is on ideas and reasoning; "notions, not notations," as the saying goes. That is not to say, however, that the book is always easy going. Mathematics takes some getting used to, and you should not be surprised if you have to pause periodically to mull over something we have said. Furthermore, things do get gradually more complex as we go along, and readers without previous mathematical experience might find some of the concluding material a bit more challenging than what came before. Even here, though, we believe we have provided enough commentary to make the central ideas comprehensible to all. In those few cases where we have elected to include some more technical material, the dense calculations can be skimmed over without losing the thread of the discussion.

The book is structured as follows: In the first chapter we examine techniques for solving Sudoku puzzles and discuss the general question of what constitutes a math problem. Chapter 2 discusses the notion of a Latin square, an object of long-standing interest to mathematicians of which Sudoku squares are a special case. Chapter 3 discusses Greco-Latin squares, which are an extension of the idea of a Latin square. Chapters 4 and 5 discuss two counting problems related to Sudoku. Specifically, we determine the total number of Sudoku squares and the total number of "fundamentally different" squares. In the course of this discussion, we cannot avoid presenting fundamental ideas from combinatorics and abstract algebra. Chapter 6 presents the problem of how one finds interesting Sudoku puzzles and places this problem within the context of search problems generally. Chapters 7 and 8 investigate connections between Sudoku, graph theory, and polynomials. Chapter 9 is an exploration of Sudoku extremes. We look for puzzles with the maximal number of vacant regions, with the minimal number of starting clues, and numerous others. The book concludes with a gallery of novel Sudoku variations. No math here, just pure solving fun! All of the puzzles presented in the text, save for a handful of exceptions that are explicitly identified, are original to this volume.

A final, bureaucratic detail. The solutions to many of the puzzles appear in the back of the book. In some cases, however, the solution to the puzzle is essential to the exposition and, therefore, has been included in the text. Wherever possible, we have placed the solution to a puzzle on a different page from the puzzle itself. Occasionally this was not possible. For that reason you may find it useful to read with an index card in hand. This will allow you to conceal portions of the page you do not wish to read immediately.

The history of math and science shows there is often great insight to be gained from the earnest consideration of trivial pursuits. Probability theory is today an indispensable tool in many branches of science, but it was born out of gambling and games of chance. In the early days of computer science and artificial intelligence, much attention was given to the relatively unimportant problem of programming a computer to play chess.

We have similar ambitions for this book. Perhaps you have tended to see Sudoku puzzles as an amusing diversion, useful only for passing the time during long airplane rides. After reading this book you will see instead a gateway into the world of mathematics. It is a far different, and more beautiful, world than you may think.

The authors would like to thank Philip Riley, whose computer prowess assisted greatly with the construction of many of the Sudoku puzzles in this book. Without Phil's work at Brainfreeze Puzzles, large portions of this book would not exist. We would also like to thank our Sudoku Master beta-tester Rebecca Field for checking all of the puzzles in the text for accuracy and playability. Finally, we would like to thank Phyllis Cohen, our editor at Oxford University Press, who was tremendously helpful and supportive throughout the writing of this book.

Taking Sudoku Seriously

1

Playing the Game

Mathematics as Applied Puzzle-Solving

What is it about puzzles that makes them so engrossing?

Imagine you are minding your own business, thinking your very practical and familiar thoughts, when someone challenges you to measure an interval of nine minutes using only a four-minute hourglass and a seven-minute hourglass. You are dismissive, perhaps, protesting you have little time for such frivolity. But the question gnaws at you, and pretty soon you are wondering what happens if you start both hourglasses going at the same time. You notice that when the four-minute glass runs out, there are three minutes left in the seven-minute glass, and then you start looking for ways to turn that to your advantage. I can time three, four, and seven minutes, you think, but how does that help me get nine? Then you are gone, your formerly practical thoughts banished until the problem is solved.

Or maybe you are presented with two bottles, one containing a liter of water, the other a liter of wine. You are told that some amount of wine is transferred to the water bottle and the resulting water-wine mixture is thoroughly stirred. Enough of this mixture is now transferred back to the wine bottle so that both bottles again possess one liter of liquid. Is there now more water in the wine bottle or wine in the water bottle? That the question seems unanswerable is part of its charm. Seriously, what can we do? Having no idea how much liquid was transferred, it would seem I can not determine either of the quantities in question. But there must be *something* I can do, as it would be a serious breach of etiquette to present a puzzle with no solution. Maybe there is something in the fact that it was pure wine that was transferred to the water bottle, as opposed to a dilute water-wine mixture that was transferred back . . . and once having started down this path, you would do well to cancel your remaining appointments for the day. (Solutions to both of these puzzles are presented in Section 1.6.)

Or maybe you are shown a 9×9 grid like this one:

7						2	9	
				3	2			6
	1					5		3
	5		1			*C*	2	
	7	4	*B*	6	*A*	8	5	
	2				3		7	
6		7					3	
1			6	9				
	9	5						2

You are challenged to fill in the vacant cells with the digits 1–9 in such a way that each row, column, and 3×3 block contains each digit exactly once.

That this is surely the most trivial of pursuits does not stop you from noticing that cell *A* has rather a lot of digits surrounding it. Certainly *A* can not be a 2 or 3, since those digits already appear in its column. Its row brings the digits 4–8 to the party, while its 3×3 block puts the kibosh on 1. This leaves 9 as the only possibility, and we happily pencil it in.

Perhaps now you notice the 2s in rows 4 and 6. They are shooting out horizontal laser beams that will burn your fingers if you try to place a 2 in the fourth or sixth rows of the central 3×3 block. But there must be a 2 *somewhere* in that block, and with the 9 filled in that leaves only cell *B*.

Suddenly all of the occupied cells are shooting out lasers! The 3s in row 6 and column 9 have the center-left 3×3 block so sliced up that the only place for its 3 is cell *C*. This is turning out to be so much fun that we had better put all else on hold until the remaining cells yield forth their secrets.

This, as you are probably aware, is an example of a Sudoku puzzle. In recent years, they have become enormously popular. Newspapers routinely present them right alongside the venerable crossword puzzle, and in-flight magazines are seldom found without them. The puzzle sections of bookstores are dominated by anthologies of Sudoku puzzles. There are countless websites devoted to Sudoku and its variants, and there are public competitions where people race to solve them.

And if there is one thing about which all of these venues agree it is that, the necessity for writing actual numbers in those little cells notwithstanding, solving a Sudoku puzzle has nothing to with mathematics. Writing in *Scientific American*, computer scientist Jean-Paul Delahaye [17] provides a blunt statement of the basic view:

> *Ironically, despite being a game of numbers, Sudoku demands not an iota of mathematics of its solvers. In fact, no operation – including addition or multiplication – helps in completing a grid, which in theory could be filled with any set of nine different symbols (letters, colors, icons and so on).*

An interesting argument, and doubtless compelling to those who regard *arithmetic* as synonymous with *mathematics.* Let us suggest, however, that there is quite a leap in going from "no addition or multiplication," to "not an iota of mathematics." And if you found anything remotely amusing in our previous discussion, then you have more of a taste for mathematics than you might realize.

1.1 MATHEMATICS AND PUZZLES

Mathematicians are professional puzzle-solvers. We are not professional arithmetic-doers. Our job is to seek out puzzles that amuse us and solve them. People pay us to do this because history shows that the earnest contemplation of amusing puzzles routinely leads to constructs of enormous practical value.

For example, in the seventeenth century, the nobleman and gambler known by his title Chevalier de Méré introduced the "Problem of Points." Imagine that two people, Alice and Bill, are taking turns flipping a coin. Alice gets a point for each heads, while Bill gets a point for each tails. The winner is the first to ten points, and the score is currently eight to seven in Alice's favor. Further assume the prize is a pot of money to which Alice and Bill have contributed equally. If the game were suspended at this point, how ought we divide the pot between Alice and Bill?

This problem came to the attention of Blaise Pascal and Pierre de Fermat, two gentlemen who rather enjoyed such puzzles. Fermat observed that since the game will end in no more than four tosses, there were only sixteen ways things can play out. We can simply list them all. We would then find that in eleven out of the sixteen cases, Alice wins, as compared to only five for Bill. Since each of these sixteen scenarios is equally likely, we should give $\frac{11}{16}$ of the pot to Alice and the rest to Bill. Pascal agreed with this division, but then one-upped Fermat by deriving general formulas for each player's chances in more general scenarios. In so doing they began a line of investigation that led to the modern theory of probability. (See the book by Rosenhouse [34] for more information and further references.)

Then there are the famous paradoxes introduced by the philosopher Zeno in the fifth century BC. One of his puzzles proposed that motion was impossible. You see, in traveling from point A to point B, you must first traverse half the distance. Doing so requires first traversing half of *that* distance or one-quarter of the total distance. No matter how small the distance, you must first traverse half of it before completing the trip. It would seem you must carry out infinitely many steps before getting anywhere, and that is why motion is impossible.

A fully satisfactory response to Zeno was not forthcoming until the seventeenth century, when Isaac Newton and Gottfried Leibniz got to wondering why, exactly, a finite distance could not be divided into infinitely many pieces. Considering a variation on Zeno's paradox, they noted that in traveling a distance of one mile you must first travel half of a mile. Then you must travel half of the remaining distance, or one-quarter of a mile. Then you must travel one-eighth of a mile and so on. Your total distance, which is one mile in this case, is then the sum of these infinitely many smaller steps. That is

$$\frac{1}{2} + \frac{1}{4} + \frac{1}{8} + \frac{1}{16} + \cdots = 1$$

But does this equation make sense? Is there a way of thinking about addition that makes plausible the notion of an infinite sum? Persist in this line of thought and you are well on your way to inventing calculus [15].

Then there is the famous story of the bridges of Königsburg. It seems that the city of Königsburg in Prussia (known today as Kaliningrad in Russia) was divided into four pieces by the Pregel River. These pieces were linked by seven bridges, as shown in this map:

The Seven Bridges of Königsburg

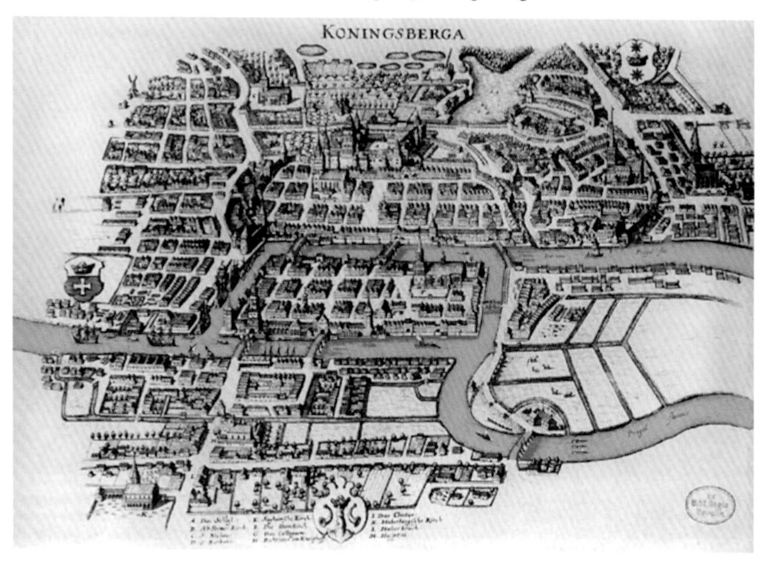

The locals had gotten to wondering whether it was possible to walk through the city in such a way that each bridge is crossed exactly once. In 1735, Leonhard Euler had the idea of representing the situation via the following abstract model: Each land mass could be thought of as a single location representable as a dot, or *vertex*. Each bridge could be thought of as a line, or *edge*, connecting two of the land mass vertices. The resulting diagram is referred to as a *graph*.

A Graph Representing the Seven Bridges of Königsburg

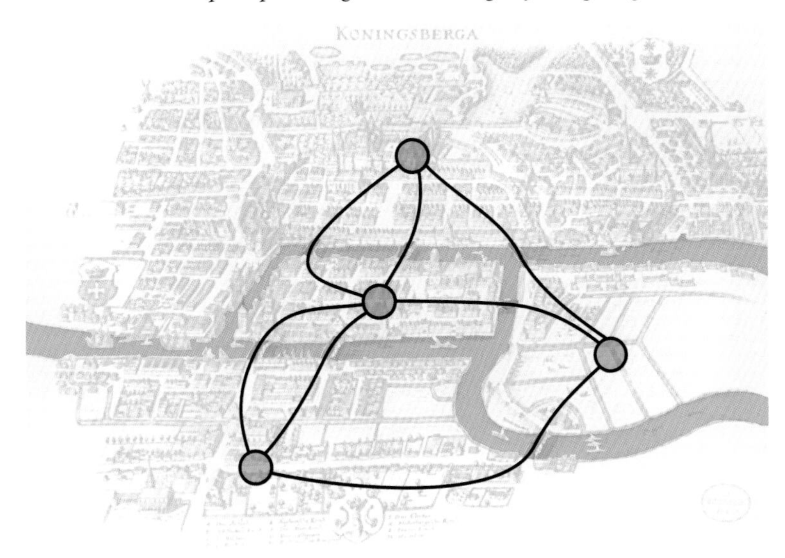

Euler noticed that every vertex has an odd number of edges coming out of it. Imagine now that you are walking through the town. Each time you first enter, and then leave, a vertex you "use up" two available edges. That means if there were a complete walk through the town, then it is only the starting and ending vertices that could have an odd number of edges. Since that is not the case here, we see that the locals will search in vain for the desired walk.

The number of edges attached to a vertex is called the *degree* of the vertex. Euler had discovered that in order for there to be a path that travels over every edge exactly once, the graph must have either exactly two vertices with odd degree (the start and end of the path), or no vertices with odd degree (in which case the path starts and ends at the same place). Such walks are now called *Eulerian paths* if they have different starting and ending points, and *Eulerian circuits* if they loop back to their starting points.

More surprisingly, it turns out that having two or zero vertices of even degree is not just a necessary condition, it is a sufficient condition as well! In other words, if a graph has either exactly two or zero odd-degree vertices, then an Eulerian path or circuit must exist.

Euler's breakthrough was one of those exceedingly clever insights that transforms a puzzle from opaque to crystal clear. It also inaugurated the branch of mathematics known as Graph Theory, which remains a going concern to this day.

Here are a couple of puzzles to whet your appetite. Note that Euler's observations can tell you if the graphs below have Eulerian circuits, but they do not tell you how to *find* these circuits. That part is up to you:

Puzzle 2: Eulerian Circuits.

An *Eulerian circuit* for a graph is a path that starts at one vertex, travels along the edges so that it visits every edge exactly once, and then returns to the original vertex. Find one in this graph:

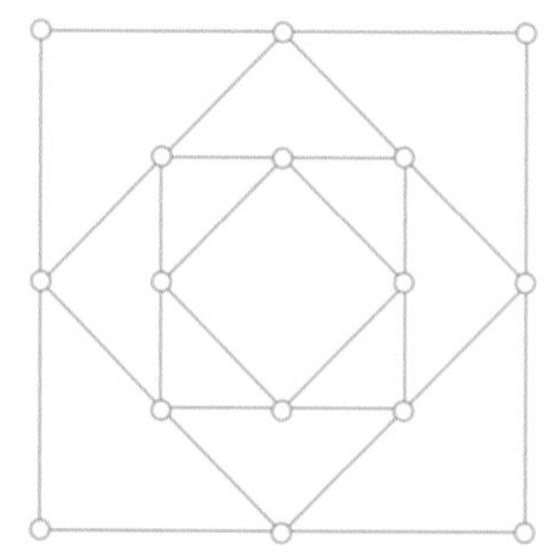

In an Eulerian circuit we traverse each edge exactly once. What if instead we want to reach every *vertex* exactly once? That is, we declare that while we are walking around the town we cannot cross a bridge multiple times, but we drop the requirement of having to traverse every edge of the graph at least once. This might come up if, for example, you were driving to a number of errands at various stores (the vertices) and did not want to retrace your steps on any of the streets (the edges).

Puzzle 3: Hamiltonian Circuits.

A *Hamiltonian circuit* on a graph is a path that starts at one vertex, travels along the edges of the graph so that it visits every *vertex* exactly once, and then returns to where it started. Individual edges can be left untraveled. Find a Hamiltonian circuit in this graph:

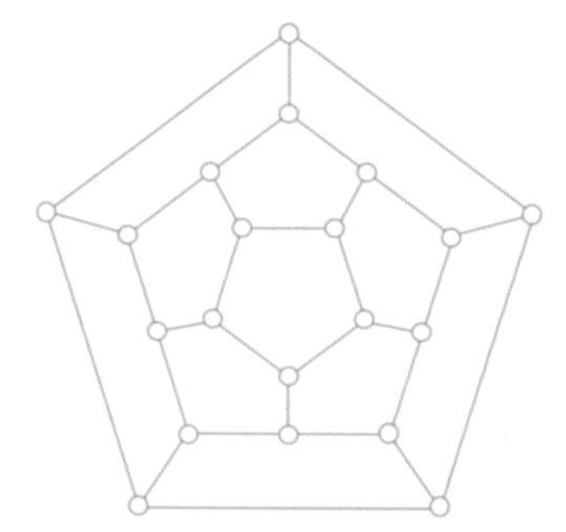

Perhaps, as anthropologist Marcel Danesi suggests, humans have an instinct for puzzles just as we have an instinct for language [16]. Maybe it is some side product of our evolution, as though natural selection says, "You can have a big brain, but only at the cost of becoming hopelessly distracted by every silly teaser to come down the road." Perhaps it is the very frivolity of puzzles that makes them so much fun. After all, were these problems important, our inability to solve them would be cause for concern, and not bemused frustration.

Whatever the reason, we will take it as given that there is something deeply satisfying in encountering opacity and, using nothing more formidable than your own intellect, producing clarity. With that in mind, let us return to our Sudoku puzzles. We will devote this and the next two sections to considering some techniques for solving them.

1.2 FORCED CELLS

We now revisit our original puzzle:

Puzzle 4: Sudoku Walkthrough.

Fill in each grid so that every row, column, and block contains each of the numbers 1–9 exactly once. We will walk through the first half of the solution in the text below.

7						2	9	
				3	2			6
	1					5		3
	5		1				2	
	7	4		6		8	5	
	2				3		7	
6		7					3	
1			6	9				
	9	5						2

By now we have developed a habit of thinking in which we do not see just eighty-one individual cells. Instead, each cell has associated to it a particular zone of the board consisting of its row, column, and 3×3 block. For example, for the cells A and C, we have the following zones:

					2			
			1					
	7	4		6	*A*	8	5	
					3			

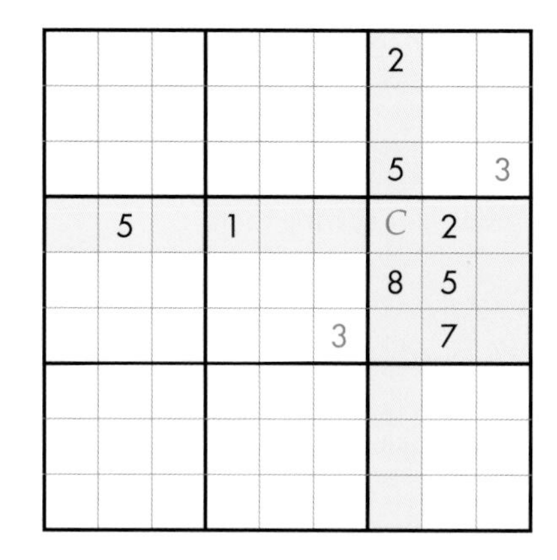

We shall refer to the cells *A* and *C* as the *generators* of their shaded zones. Our first solving technique used the fact that the digit found in the generating cell must be different from every other digit in the zone. We determined the value of *A* by inspecting its zone and noting that 9 was the only absent digit.

We applied a different technique to *C*. Merely inspecting its zone was inadequate in this case since 3, 4, 6, and 9 are all currently missing. Thus, instead of choosing a cell and asking which value it could contain, we selected a particular region (*C*'s 3 × 3 block), noted that it must contain a 3, and asked which of its vacant cells was available for that purpose. The red 3s shown outside of *C*'s zone force *C* to be 3.

These examples suggest a general starting point for solving Sudoku puzzles. Examine the zone of each vacant cell and pencil in all of its candidate values. Having done so, look for the following types of *forced cells*:

1. *One-choice*: A cell that contains only one candidate value.
2. *One-place*: A region (row, column, or block) that has only one cell available for a given number.

Many easy puzzles can be solved with nothing more than forced cell techniques. Let us try applying them to our current puzzle.

Below left we see the zones for all the 1s that are currently placed. Notice that in the leftmost middle block there is only one place where a 1 could go (indicated by the vacant circle). By the method of one-place we can enter a 1 in that cell. This permits some further shading, which produces another block – the rightmost middle one – in which there is now only one place for a 1.

7						2	9	
				3	2			6
	(1)					5		3
	5		(1)			3	2	
	7	4		6	9	8	5	
	2	○			3		7	
6		7					3	
(1)			6	9				
	9	5						2

7						2	9	
				3	2			6
	(1)					5		3
	5		(1)			3	2	
	7	4		6	9	8	5	○
	2	(1)			3		7	
6		7					3	
(1)			6	9				
	9	5						2

With nothing more than repeated scanning for one-place situations and filling in any obvious one-choice cells, you can fill in all of the cells with asterisks below. Have a go at it and join us on the other side:

7	*	*				2	9	*
*				3	2	*	*	6
*	1				*	5	*	3
	5	*	1			3	2	
*	7	4	*	6	9	8	5	1
	2	1			3	*	7	
6		7		*		*	3	*
1	*	*	6	9	*	*	*	*
	9	5	*			*	*	2

What now? We could keep going using just one-place and one-choice, but let's look at a new technique.

1.3 TWINS

Our Sudoku board now looks like this:

7	6	3				2	9	8
5				3	2	7	1	6
2	1				6	5	4	3
	5	6	1			3	2	
3	7	4	2	6	9	8	5	1
	2	1			3	6	7	
6	48	7	48	2	148	9	3	5
1	3	2	6	9	5	4	8	7
	9	5	3			1	6	2

The tiny red values in the seventh row are the candidates for these cells. That is, they are the numbers that are not immediately eliminated by considering the zones of the other cells on the board.

Having run out of forced cells, we will have to try something a bit more clever. Instead of looking for cells with only one candidate value, perhaps we should go looking for cells with two. For example, the first two open cells in the seventh row cannot possibly contain any values other than 4 or 8.

These two cells are an example of a *twin*. They are a pair of cells in the same region having the same two candidate values. Twins are potentially very helpful. In the present case, for example, we can be absolutely certain that between these two cells, one of them contains a 4 while the other contains an 8. If the first one is 4 then the second must be 8, and vice versa.

Look now at the third open cell in the seventh row. Examining zones shows that its only candidates are 1, 4, and 8. But 4 and 8 must already appear in the first two cells of the seventh row, in a currently unknown order. This tells us that the third open cell in the seventh row must in fact contain a 1. Remarkable! Even without specifying the order of the 4 and 8 in the first two cells, we can determine the value of the third.

Filling in the 1 we just identified, plus the two other circled cells we get by scanning anew for one-place situations, we get the board below left. Repeating the method of twins now helps us in the fourth column. Two of its open cells contain only 4 and 8 as candidates. This pair of twins tells us that no other cell in the fourth column can contain a 4 or an 8. This forces the topmost open cell in that column to be 9, which in turn forces the second open cell in that column to be a 7, as shown below right:

7	6	3	(5)	(1)		2	9	8
5			489	3	2	7	1	6
2	1		789		6	5	4	3
	5	6	1			3	2	
3	7	4	2	6	9	8	5	1
	2	1	48		3	6	7	
6		7	48	2	(1)	9	3	5
1	3	2	6	9	5	4	8	7
	9	5	3			1	6	2

7	6	3	5	1		2	9	8
5			(9)	3	2	7	1	6
2	1		(7)		6	5	4	3
	5	6	1			3	2	
3	7	4	2	6	9	8	5	1
	2	1			3	6	7	
6		7		2	1	9	3	5
1	3	2	6	9	5	4	8	7
	9	5	3			1	6	2

The rest of the puzzle should now fall into place rather quickly. Finish it up and join us in the next section. If you get stuck or want to check your answer, the solution is in the back of the book.

1.4 *X*-WINGS

The previous puzzle was solvable using only forced cells and twins. Sadly, sometimes more is required. Consider, for example, the following puzzle:

Puzzle 5: Harder Sudoku.

Fill in each cell so that every row, column, and block contains each of the numbers 1–9 exactly once. We will walk through the trickier parts of the solution in the text below.

				9				
					7			1
				4		5	7	
8			5			1	4	
	2	7				3	9	
	3	4			2			8
	6	9		5				
2			6					
				1				

The cells with asterisks as shown below left are not so hard to fill in, so let's start with those. The result is shown below right. Cover up the right side and see if you can get there yourself!

				9	*			
				*	7			1
		*		4	*	5	7	
8	*	*	5	*	*	1	4	*
*	2	7	*	*	*	3	9	*
*	3	4	*	*	2	*	*	8
	6	9		5			*	
2			6	*			*	
				1				

				9	5			
				2	7			1
		2		4	6	5	7	
8	9	6	5	7	3	1	4	2
5	2	7	4	8	1	3	9	6
1	3	4	9	6	2	7	5	8
	6	9		5			1	
2			6	3			8	
				1				

Things get tricky now, but here is one way to proceed. As shown below left, a set of twins in the third row allows us to eliminate a 3 from consideration in the fourth cell of that row. Now look below right. What worked for pairs of numbers works just as well for triples. The first three cells of the fourth column have candidate values entirely among 1, 3, and 8, implying that no other cell in that column can contain those values. This allows us to remove the candidate 8s from the last two open cells in the fourth column. If we now look at the circled cell, we see that it is the only one in its row that can contain an 8.

				9	5			
				2	7			1
39	18	2	1~~3~~8	4	6	5	7	39
8	9	6	5	7	3	1	4	2
5	2	7	4	8	1	3	9	6
1	3	4	9	6	2	7	5	8
	6	9		5			1	
2			6	3			8	
				1				

			138	9	5			
			38	2	7			1
		2	18	4	6	5	7	
8	9	6	5	7	3	1	4	2
5	2	7	4	8	1	3	9	6
1	3	4	9	6	2	7	5	8
	6	9	27~~8~~	5	○		1	
2			6	3			8	
			27~~8~~	1				

This brings us to the strategy known as *X-Wings*. Look at the possible candidates in the third and seventh rows shown below left. In each of these rows the 3 can only appear in the first or last cell. That means the 3 in the first column appears either in row 3 or row 7, and likewise in the last column. Therefore, as shown below right, we can eliminate all candidate 3s from the remaining cells in those columns. This allows us to enter a 4 in the northeast corner of the puzzle!

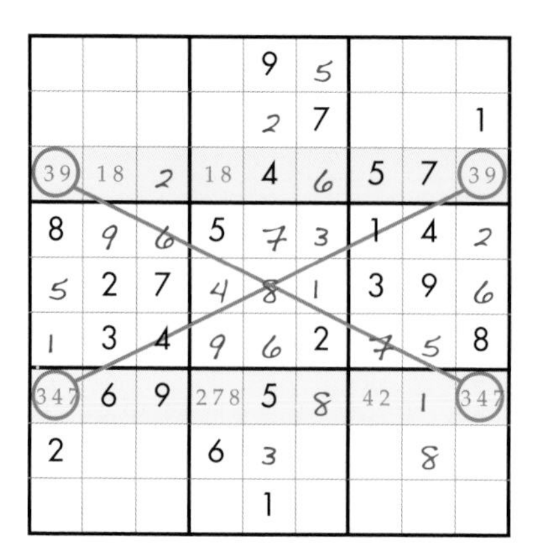

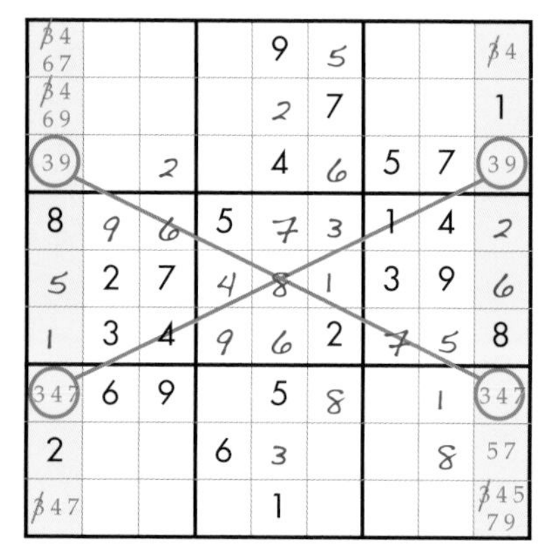

Notice that the four cells with green circles form the corners of a rectangle. Given a 3 in one corner, the diagonally opposite corner is forced to contain a 3

as well. Since the relationships between pairs of opposite corners in the rectangle can be depicted by a large X, this structure is known as an X-Wing.

Alas, even after all that work the puzzle still will not fall. It is time to pull out the big guns.

1.5 ARIADNE'S THREAD

We are working hard for each cell, but it will soon be worth it. Look at the labeled cells below left and the candidate values for those cells shown below right. We see that the cell labeled A contains either a 3 or a 9, but we do not know which. Our strategy will be to make a preliminary guess as to which value is correct. We will then follow the logical consequences of that decision, in the hopes of thereby gaining some insight regarding the vacant cells.

				9	5	*	*	4
				2	7	F	*	1
		2		4	6	5	7	A
8	9	6	5	7	3	1	4	2
5	2	7	4	8	1	3	9	6
1	3	4	9	6	2	7	5	8
	6	9	C	5	8	D	1	B
2			6	3		E	8	
				1				

				9	5	268	236	4
				2	7	689	36	1
		2		4	6	5	7	39
8	9	6	5	7	3	1	4	2
5	2	7	4	8	1	3	9	6
1	3	4	9	6	2	7	5	8
	6	9	27	5	8	24	1	37
2			6	3		49	8	
				1				

Let us try placing a 3 in cell A. That forces cell B to contain a 7. Cell C is then forced to contain a 2, which forces cell D to be a 4. Then cell E contains a 9. As fun as this is, however, our luck is about to run out. For now we must inquire as to the location of the 9 in the upper right block. Having placed a 3 in cell A, and having been forced thereby to place a 9 in cell E, we are entirely out of options. It seems that our experiment of placing a 3 in cell A has led to a contradiction. But since that forces us to place a 9 in that cell we see that, after all our hard work, another cell has fallen.

In his essay on solving Sudoku puzzles [30], Michael Mepham refers to this style of thought as "Ariadne's thread." The reference is to Greek mythology. It seems that Ariadne's lover, Theseus, had entered the labyrinth of King Milos with the intention of killing the dreaded Minotaur. To keep him from getting lost, Ariadne gave Theseus a long, silken thread. Theseus unrolled the thread as he proceeded through the maze. Then, upon hitting a dead end, he could backtrack to the most recent fork and take a different path.

That is precisely what we have done. We followed the path of placing a 3 in cell A until it led us to a dead end. We rectified our error by backtracking to the point of our fallacious assumption and replaced it with a more reasonable choice.

You might object that Ariadne's Thread hardly counts as a solving technique, since it seems tantamount to guessing. We would suggest, however, that this is not the best way of looking at things. In a proper Sudoku puzzle, the value in each cell is logically determined by the placement of the starting clues. A 'solving technique' is any method that aids you in discerning the relevant deductions. In some cases, as with a forced cell or a twin, the logic is straightforward and easy to see. In others, like a triple or an X-Wing, more subtle reasoning is required. Regardless, in every case you are asking yourself, "What are the logical consequences of placing this digit in this cell?"'

So it is with Ariadne's Thread. It is not that this technique involves guessing, whereas our other techniques do not. It differs from the other techniques only in the complexity of the deductions needed to make it work. This is not surprising. After all, Ariadne's Thread is the technique to which you resort after the simpler methods have proven inadequate. In some especially difficult puzzles, the logical chain might be of such length and complexity that it defeats the abilities of all but the most skillful solvers. For all of that, however, it is not fundamentally different from our other solving techniques, and it is not comparable to outright guessing.

We have now forced our way through the roadblocks of this puzzle, and the rest falls in line fairly easily. You can finish up and join us after the jump.

1.6 ARE WE DOING MATH YET?

Suppose there is a barn containing cows and chickens, 50 animals in all. We notice that there are a total of 144 feet on the ground. Keeping in mind that chickens have 2 feet while cows have 4, can you determine the number of cows and chickens?

No doubt you took an algebra class at some point in your life, and if you did you might come up with an argument like this: Let x denote the number of cows. Then $50 - x$ denotes the number of chickens. Then the number of cow feet is $4x$ while the number of chicken feet is $2(50 - x)$. Since the total number of legs is 144, we have

$$\begin{aligned} 2(50 - x) + 4x &= 144 \\ 100 + 2x &= 144 \\ x &= 22. \end{aligned}$$

There must be 22 cows, and therefore 28 chickens, in the barn.

Now *that's* a math problem! We used algebra and everything.

However, if using algebra were the criterion, then this would cease to be a math problem as soon as someone thinks of telling the cows to stand on their hind legs. There would then be fifty animals, each with two feet on the ground, for a total of one hundred feet. That means there are forty-four feet in the air. Since each cow has two feet in the air, there must be twenty-two cows. Simple as that.

Attention, cows, please stand up

Yes, you might say, that is terribly clever. But we used arithmetic so it is still math.

Then what about the hourglass problem from the chapter's preamble? We were asked to time a period of nine minutes using only a four-minute and a seven-minute hourglass. Here is one possible solution. For convenience, we will refer to the four-minute glass as F and the seven-minute glass as S. Begin by flipping over both of them. After four minutes, F is empty while S still has three minutes to go. Now flip over F. Three minutes later, after a total of seven minutes have elapsed, S is empty while F has one minute left. Flip over S. One minute later F is empty, while S has one minute of sand in its base. Now flip S again. When it runs out exactly nine minutes have elapsed. Here it is in pictures:

Clever hourglass flipping to measure nine minutes

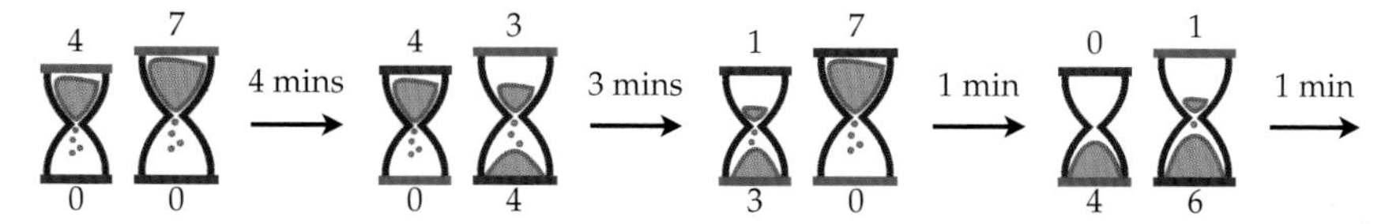

Was that a math problem? No arithmetic, really, just basic counting and a bit of cleverness. But you might still argue that numbers were involved, so it is still math.

Then what about the water and wine problem? Recall that we had a liter of water and a liter of wine. Some of the wine was transferred to the water bottle and mixed in thoroughly. Then enough of the water/wine mixture was transferred back to the wine bottle so that both containers again contained one liter of liquid. Is there now more wine in the water bottle or water in the wine bottle?

One of the charming aspects of this problem is that it is typical to go from complete befuddlement to perfect understanding in an instant. There is no middle ground where you sort of see what is going on. The crucial insight is that while some unknown quantity of wine has left the wine bottle, at the end of the problem the wine bottle contains precisely as much liquid as it did at the start. If you imagine that the wine and water become separated in each of the glasses, the final situation must look something like this:

Water in the wine, wine in the water

Whatever wine has left the wine glass has been replaced by precisely the same quantity of water. Meanwhile, every last drop of the missing wine is now residing in the water glass. We conclude that the quantity of wine in the water bottle is identical to the quantity of water in the wine bottle.

No algebra, no arithmetic, no counting, no numbers. Yet unambiguously a math problem.

Mathematics, you see, is really characterized by the use of deductive logic. If the problem you are contemplating can be solved solely through deductive logic, then you are working on a math problem.

Which is not to say that mathematicians are the only ones who use logical deduction. Logic plays a central role in every branch of science, and many other disciplines besides. Indeed, philosophers often refer to the Hypothetico-Deductive model of scientific practice, in which scientists formulate a hypothesis, deduce its logical consequences, and then design an experiment to test if the conclusions are seen to hold. Still, while scientists use logic as a tool in their work, their problems cannot be solved by logic alone. No amount of armchair cogitation will establish the workings of a cell or the nature of electricity.

Nor are we saying that mathematicians are perfect little reasoning machines, coldly grinding out all of the logical implications inherent in a set of axioms. Not at all. It is true that we do not regard a problem as solved until there is a sound chain of deductions leading from the given information to the resolution of the question, but that final argument is the end result of mathematical work, not the sum total of it. Those elegant proofs and dense calculations you find in the textbooks give you no inkling of the road that was traveled in their construction. That road typically features many hours of experimentation, imagination, intuitive leaps, conversations with students and other mathematicians, and, if the problem is at all interesting, many hours of frustrated wall-staring.

If it is the use of deductive logic, as opposed to arithmetic or algebra or all the rest, that characterizes a math problem, then why do math classes spend so much time on tedious algorithms for computation, or the seemingly arbitrary rules for symbol manipulation? Why the close association of mathematics with numbers, shapes, functions and the numerous other abstract constructions that cause so much frustration?

It is simply because those are the sorts of objects to which deductive reasoning best applies. Unique among the sciences, mathematicians have the satisfaction of knowing that when they solve a problem it stays solved. Sound deductive arguments have a permanence and certainty about them that other ideas in science lack. This permanence is an endearing feature of our discipline, but we pay a heavy price. That price is our exile to planet Abstraction, which is a very different place from planet Earth. That the two planets nonetheless have much to say to one another is one of those delightful facts about the world that philosophers still have not adequately explained.

At any rate, for most of the time you spent learning the techniques of algebra and calculus, you were not really doing mathematics at all. Instead you were learning, and hopefully mastering, a set of techniques that are routinely useful in solving problems. You must learn to skate if you are going to play hockey, but all the skillful skating in the world does not make you a hockey player. You must learn to use a band saw and a drill press if you are going to work with wood, but there is far more to carpentry than using a few tools. So it is with mathematics. Algebra and the rest are tools that we use for solving problems, but it is the problems themselves that form the core of our discipline.

Which brings us back to Sudoku puzzles. They are solved through pure logic. Therefore, they are math problems. Show us someone who says otherwise, and we will show you someone with too narrow a view of our discipline.

1.7 TRIPLETS, SWORDFISH, AND THE ART OF GENERALIZATION

All mathematical theorems ultimately say the same thing. They say that if you define certain terms just so, and if you grant certain assumptions, then certain conclusions follow as a matter of logic.

It stands to reason that the fewer assumptions we have, the more useful the theorem is likely to be. That is why mathematicians, after proving a theorem, will ask themselves whether all of their assumptions were truly necessary. Proving a theorem is nice, but showing that a known theorem is just a special case of something more general is even nicer.

For example, you are probably familiar with the Pythagorean theorem. In any right triangle, the lengths of the sides and hypotenuse are related by the equation:

For a right triangle, $a^2 + b^2 = c^2$

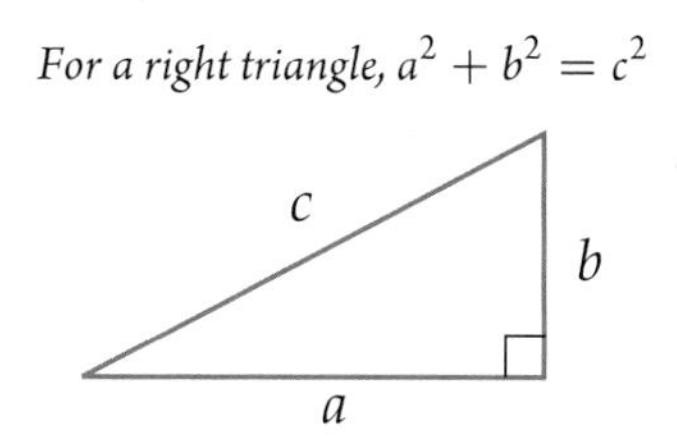

It is a nice theorem, but it only applies to right triangles. There is a more general statement, known as the law of cosines, that applies to any sort of triangle:

For any triangle, $c^2 = a^2 + b^2 - 2ab\cos\theta$

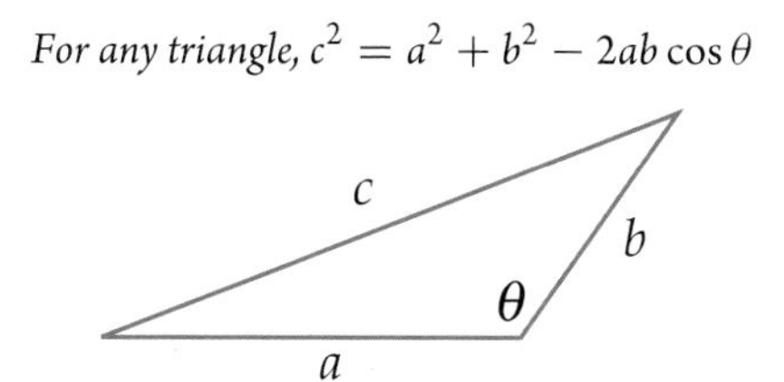

Of course, if θ is a right angle then its cosine is 0, and this equation reduces to the familiar Pythagorean theorem.

We can apply this sort of thinking to our Sudoku solving techniques. Consider, for example, the method of twins. We noted that if two cells in the same Sudoku region have the same two candidate values, then we can be certain that those two values go in those two cells. Since that implies those candidate values cannot appear elsewhere within that region, we have discerned a very useful piece of information.

Is there anything special about the number 2 in that example? That is, does it only work for the case of two values in two vacant cells? Certainly not! As we saw in our solution to Puzzle 5, three candidate values in three vacant cells would work just as well.

What about four? For example, suppose you mark out all the possible candidates for the open cells in a given 3×3 block and notice that four of the open cells have the following candidates: 145, 148, 1458, 458. The numbers 1, 4, 5, and 8 must appear, in some arrangement, in those four cells. That means that any *other* open cell in that block cannot contain those four numbers. We can cross out those numbers as candidates in all of the other open cells of the block.

You can see where this is going. We can now formulate a general rule: If n cells within a single Sudoku region have candidate values drawn from the same n-element set, then those n values must appear in those n cells.

How about the X-Wing? Can we generalize that?

The idea was that if two rows (or two columns) each have only two vacant cells that can contain a particular digit, and if those four cells form the corners of a rectangle, then you can be certain that the digit in question appears in one of the pairs of diagonally opposite cells. Said another way, if two rows each only have two cells that can contain a particular digit, and if all of those cells lie within two particular columns, then we can eliminate all possibilities for that digit in the other open cells in those two particular columns. The situation is similar if rows and columns are interchanged.

We can also do this with *three* rows or *three* columns. For example, if each of three rows has three or fewer cells that can contain a particular digit, and if all of those cells happen to lie within three particular columns, then we conclude that the digit does not appear in any *other* cells in those columns. Instead of the corners of a rectangle, which form a 2×2 lattice of cells, we are now dealing with a 3×3 lattice of cells. This technique is commonly known as *Swordfish*, probably because the cells in the 3×3 lattice that end up being used in this technique sometimes have a fishy shape.

We can generalize further with a 4×4 lattice, where we have four rows each containing four or fewer possible locations for a digit, and all of these locations lie within four particular columns. This configuration is known to some as *Squirmbag*, which if you think about it just means that the pattern within the 4×4 lattice doesn't really look like anything at all. We could now generalize to n rows with a digit appearing in only n possible intersecting columns.

Sudoku not a math problem? Nonsense! It is a perfect model for math in the small. Even better, the connections between math and Sudoku extend far beyond the mechanics of solving the puzzles. Exploring those connections will be the main focus of the ensuing chapters.

1.8 STARTING OVER AGAIN

One does not have to name or even identify fancy techniques to play Sudoku. With a bit of experience, you inevitably stumble upon the techniques we have mentioned, along with several others besides. That is the whole fun of Sudoku – developing your own ways of navigating the maze.

Alas, after working through a large number of puzzles the experience can become a bit stale. The solution is to move on to new Sudoku variations that force you to develop new techniques. We could consider puzzles that have additional regions beyond the usual rows, columns, and 3×3 blocks. For these puzzles, the concepts of forced cells, twins, triples, X-Wings, and so on have to be generalized to take the additional regions into account. Here are two such puzzles for you to play with. In the first, we add the requirement that the two main diagonals of the square also contain the numbers 1–9 exactly once. In the second, we add four new 3×3 blocks as regions that must contain 1–9 exactly once.

Puzzle 6: Sudoku X.

Fill in the grid so that every row, column, block, and main diagonal contains each of the numbers 1–9 exactly once.

	2							5
3		6						2
	4		2				9	
		3		8		9		
			5	1	9			
		9		6		7		
	7				1		8	
6						5		7
8							2	

Puzzle 7: Four Square Sudoku.

Fill in the grid so that every row, column, and block, and each of the four additional shaded blocks contains each of the numbers 1–9 exactly once.

5	4						6	3
6							5	4
		1				2		
					3			
			6	9	2			
			7					
		6				7		
3	1							9
2	5						3	1

If you need a hint for the puzzle above, try investigating the 6s; you should be able to place them all without too much difficulty. Four Square Sudoku is a relatively popular variation and has recently been investigated in a nice paper by Michel [31].

Not enough of a challenge? We can do worse. Instead of adding regions which force us to generalize our previous solving strategies, let us change the rules themselves. That will force us to make up entirely new strategies.

One way of doing this involves allowing certain symbols to be repeated in the various Sudoku regions. In Puzzle 8, each region contains *three* stars. In Puzzle 9 we have three different symbols that each appear *twice* in each region. Tread carefully!

Puzzle 8: Three Star Sudoku.

Fill in the grid so that every row, column, and block contains each of the numbers 1–6 exactly once as well as exactly three stars.

		6			1			
★			5		4	6		
	★			6		★	2	
		★					3	6
1			★		★			5
4	2					★		
	3	★		1			★	
		5	6		2			★
			4			5		

Puzzle 9: Double Trouble Sudoku.

Fill in the grid so that every row, column, and block contains each of the odd numbers 1, 3, and 5 exactly once, and each of the even numbers 2, 4, and 6 exactly twice.

				4				2
4	4				5	6		
	2	2				1	5	
			6	6			1	
3								2
	2			4	4			
	1	3				2	2	
		5	1				6	6
4				5				

2

Latin Squares

What Do Mathematicians Do?

As we have seen, a *Sudoku square* is a 9×9 grid in which each row, column, and 3×3 block contains the digits 1–9 exactly once. A *Sudoku puzzle* is a partially filled-in grid that has only one completion to a Sudoku square. Each Sudoku square has many possible Sudoku puzzles. Puzzles 10 and 11 exhibit two different Sudoku puzzles whose solution is the same Sudoku square.

Puzzles 10 and 11: Sudoku Brothers.

Fill in each grid so that every row, column, and block contains each of the numbers 1–9 exactly once. The first puzzle is an easy, level 1 puzzle. The second puzzle is more difficult, say level 4 out of 5; you may even find that you need Ariadne's Thread. Of course, since both puzzles have the same solution, you should not look at one puzzle while solving the other! The solution is in the text below.

1			3					7
	7			4		8	1	
		8	1			5	6	
		5	2					6
	2						8	
7					8	2		
	5	1			9	7		
	8	6		3			4	
9					4			2

							2	7
	7		9					3
		8	1			5		
	4	5			1			
				5				
			4			2	9	
		1			9	7		
2					7		4	
9	3							

Both Puzzle 10 and Puzzle 11 have the following Sudoku square for their solution. Take a moment to consider its structure. Notice how the individual digits are spread throughout the grid. None of the 1s, for example, appear in the same row, column, or 3×3 block. Likewise for the other digits.

A completed Sudoku square

1	6	4	3	8	5	9	2	7
5	7	2	9	4	6	8	1	3
3	9	8	1	7	2	5	6	4
8	4	5	2	9	1	3	7	6
6	2	9	7	5	3	4	8	1
7	1	3	4	6	8	2	9	5
4	5	1	6	2	9	7	3	8
2	8	6	5	3	7	1	4	9
9	3	7	8	1	4	6	5	2

Suppose we were to remove the block condition on a Sudoku square. We would then seek a square in which each of the digits from 1–9 appears exactly once in each row and column. Mathematicians have been studying such objects for centuries and refer to them as *Latin squares*. (The name originates from the use of Latin letters, instead of digits, in the first serious studies of these objects).

Let us give a more precise definition. Imagine that you have a collection of n distinct symbols. The first n letters of the alphabet perhaps (assuming that n is not larger than 26), or the first n positive whole numbers. A Latin square of order n is then an $n \times n$ array in which every row and column contains each of the n symbols exactly once. We refer to the number n as the *order* of the Latin square. A Sudoku square is then seen to be a Latin square of order 9 with an extra condition regarding the 3×3 blocks.

2.1 DO LATIN SQUARES EXIST?

There is an old saying that you cannot define something into existence. Declaring a unicorn to be a horse with a horn on its head in no way implies such creatures exist. In light of this, upon seeing a new definition, a mathematician will wonder whether there is anything that satisfies it.

Actually, in the case of Latin squares we have already answered that question. In the previous section, we gave an explicit example of a Sudoku square, which is a Latin square of order 9. Apparently we have not carelessly defined something that is logically impossible. We can also have Latin squares that fail the 'block condition' of Sudoku. For example, the partially filled-in grids in Puzzles 12 and 13 uniquely determine Latin squares which are *not* valid Sudoku squares. Sudoku veterans, take a deep breath, in the next puzzle you are going to have to give up all your block-related solving techniques!

Puzzles 12 and 13: Latin-Doku.

Fill in each grid so that every row and column contains each of the numbers 1–9 exactly once. The first puzzle is fairly easy and its solution is in the text below. The second puzzle is more of a challenge; its solution is at the back of the book.

	3			7	9		6	5
7	6	1		9		5	3	
	5		7			1		
	9	2	8		4			
6	8			4			2	7
			9		6	2	5	
		5			1		9	
	2	9		5		4	1	3
4	7		5	1			8	

		6	5	1	2			9
9		8					4	6
		2						
	6	1	4	9			3	
6			1	4	5			7
	2			5	7	4	1	
						2		
2	4					7		8
1			2	7	9	8		

Have you noticed that Latin-Doku puzzles seem to require more starting clues than their Sudoku cousins? There are fewer restrictions on the placement of digits in a Latin square, implying that each starting clue in Latin-Doku is less informative than its Sudoku counterpart. Any starting clue in a Sudoku puzzle tells you something about twenty other cells in the grid. For example, if the digit 5 appears as an intial clue, then you know there is not a 5 in the eight other cells in the same row, in the eight other cells in the same column, and in the four cells in the same 3×3 block that are not in the same row and column. A starting clue in Latin-Doku, by contrast, only conveys information about sixteen other cells. Since each clue is less informative, more starting clues are needed to ensure a unique solution.

Latin-Doku does not seem to be as enjoyable as Sudoku. The interactions of the 3×3 blocks with the other regions add an important dimension to the Sudoku solving experience. You might wonder, though, which regions other than 3×3 blocks could serve as acceptable regions in a Sudoku-like puzzle. We saw some possibilities in Chapter 1, and we shall encounter others as we go along.

The first Latin-Doku puzzle above has the following Latin square as its solution:

A completed Latin square

1	3	4	2	7	9	8	6	5
7	6	1	4	9	8	5	3	2
9	5	8	7	2	3	1	4	6
5	9	2	8	3	4	6	7	1
6	8	3	1	4	5	9	2	7
3	1	7	9	8	6	2	5	4
2	4	5	3	6	1	7	9	9
8	2	9	6	5	7	4	1	3
4	7	6	5	1	2	3	8	9

The row and column conditions of Latin squares force the nine occurrences of each digit to be spread out roughly evenly over the board. If we also imposed the 3×3 block condition of Sudoku, then the occurrences of each digit would be even more spread apart. It is as if each digit in a Sudoku puzzle has a zone of isolation associated with it, forcing all other occurrences of the same digit to go elsewhere. Our next two puzzles consider a condition that forces an even higher degree of spreading among the numbers.

Puzzles 14 and 15: Bomb Sudoku.

Fill in each grid so that every row and column contains each of the numbers 1–9 exactly once. In addition, no two adjacent cells (including diagonally) can contain the same number. Think of each number as a bomb that explodes into its eight surrounding cells; you must place the numbers so that no number can "attack" one of its own. The circle in the center of the board is there to remind you of the bomb condition. Watch out; the bomb condition is easy to lose track of, and the second puzzle is quite difficult.

							6	7
	1	2		3			9	8
	4	3						
5			8		1			3
	6	4		5		9	8	
4			1		6			5
						1	2	
6	7			8		4	3	
9	8							

5	2						7	
3	1			5			8	9
			6	3	7			
	5	6		4		8	2	
			5	7	3			
1	3			6			4	5
	4						1	2

2.2 CONSTRUCTING LATIN SQUARES OF ANY SIZE

We have now firmly established that 9×9 Latin squares exist. Perhaps, though, there is something special about the number 9. Are there Latin squares of order 10? 11? More generally, given an arbitrary positive integer n, is there necessarily a Latin square of order n?

Try it yourself. Construct a 10×10 Latin square using the numbers 1–10 as your symbols. Since all ten of these numbers must appear in the first row in *some* order, we might as well assume they occur in the usual numerical order. Given that first row, can you fill in the remaining nine rows to make a Latin square of order 10?

Puzzles 16: Constructing a Latin Square of Order 10.

Fill in the grid so that every row and column contains each of the numbers 1–10 exactly once. There are many possible ways of doing this. One possible solution is discussed in the text below.

We cannot answer the existence question for Latin squares of every order n simply by giving examples. There are infinitely many choices for the value of n. No matter how many specific examples we give, we shall still be left with infinitely many orders unrepresented. There are other worries. What happens if n is very large, like a billion? Such a square would contain 10^{18} entries (that is, one billion squared), which would make it effectively impossible to write down.

What is needed is a specific procedure that will allow us to produce Latin squares of whatever order is requested. A procedure, moreover, that can be shown to produce Latin squares of any arbitrary order without having to go through the tedium of actually writing it down.

How do we find such a procedure? We begin with trial and error. Let us try to construct specific examples in the hopes that the experience we gain will point us toward a general solution. Show us a mathematician who has solved a big problem, and we will show you someone who gamely groped in the dark for a while, experimenting and messing around in search of a good idea. It has been our experience as math teachers that students get nervous when a problem requires anything more than the mechanical application of a prepackaged algorithm. But real mathematics is all about getting stuck. It is the period of frustrated groping that makes the eventual solution so satisfying.

There is, in fact, a simple, systematic way of constructing Latin squares of any order. You may have just discovered it when solving Puzzle 16. Let us try first to create Latin squares of relatively small orders. With a little trial and error we quickly arrive at the squares of order 2, 3, 4 and 5 shown below.

Latin squares constructed with an obvious pattern

1	2
2	1

1	2	3
2	3	1
3	1	2

1	2	3	4
2	3	4	1
3	4	1	2
4	1	2	3

1	2	3	4	5
2	3	4	5	1
3	4	5	1	2
4	5	1	2	3
5	1	2	3	4

The pattern is clear. In each case, the first line has its entries in numerical order. Each subsequent line is obtained from the one before by shifting every number one place to the left, with the first entry in each row being moved to the end. We do this until producing one more row would return us to where we started. In the four cases above, we were successful, and it seems reasonable to suppose our procedure would work as well in other cases. One possible solution to Puzzle 16 involves cycling the numbers 1–10 in just this fashion.

A mathematician, however, would not yet be satisfied. "Seems reasonable" is too unstable a foundation for future work. What if we have overlooked a subtle point? If we have, then anything built on this foundation will be of dubious validity. We need a clear explanation for why this procedure works. How can we prove that our trick of shifting the rows always produces a Latin square? We need to verify that, in the square resulting from our procedure, each number appears exactly once in every row and column.

The rows seem unproblematic. It was part of our construction that the first row contains each number exactly once. Since every subsequent row is obtained from the first by a simple shift of the numbers, we need not worry that some number will suddenly appear twice in any of the rows.

Likewise, the columns do not pose a problem. Notice that in each of our examples, the kth column reads the same top to bottom as the kth row reads left to right. This is a consequence of our shift technique.

This all seems very plausible, but mathematicians go one step further. At this point, we pause to write down a formal proof of our result. The idea is to distill things down to their essence and to give the most precise statement we can of what we have learned. We will state our result with care and precision, so that a reader not privy to our earlier discussion will understand what we have accomplished. In the present case, things might look like this:

Theorem 1
Let n be a positive integer. Let L be the $n \times n$ matrix whose kth row, for $1 \le k \le n$, reads from left to right as $(k, k+1, \ldots, n, 1, 2, \ldots, k-1)$. Then L is a Latin square of order n.

Proof
We need to show that each of the numbers $1, 2, \ldots, n$ appears exactly once in each row and column. It is clear from our construction that this must be true in each row. To prove that we have this property for each column, consider the jth column, where $1 \leq j \leq n$. The topmost entry in this column is the jth entry of the first row of the matrix. Since the first row is given by $(1, 2, \ldots, n)$, this jth entry is the number j. By construction, each row has entries that are shifted one to the right from the previous row; therefore, the second entry of the jth column, which is the jth entry of the second row, must be $j + 1$. Continuing this pattern, we see that the jth column reads from top to bottom as

$$(j, j + 1, \ldots, n, 1, 2, \ldots, j - 1).$$

This clearly forces each number to appear exactly once in each column, as desired. □

In pondering this bit of technical bravado, we come to one of the first great truths of mathematics: A simple and straightforward idea can be made to seem very complicated when written with a high level of precision.

This presents a challenge to those of us who teach mathematics. Textbooks tend to include only the theorems and proofs, which are presented in a style that stresses efficiency over clarity. The intuition and explorations that preceded the proof are often omitted. The illusion is thus created that mathematicians summon forth theorems from some reservoir to which others may have been denied access. The reality, by contrast, is that the theorem and the proof are the end of the process, not the beginning, just as a finished novel is the end result of a series of more rudimentary drafts.

Theorem is one of those words you do not see much outside of mathematics. Its origin, though, nicely illustrates the point we are making. The word comes from Greek where it meant "to look at." The word *theater,* which is a place you go to watch a dramatic production, has the same root. A theorem is the end result of prolonged "looking at."

The precision and formal proofs are necessary as a check on our intuition, but they are not replacements for it. In learning mathematics, you need two tracks going simultaneously. One track is the intuition and the concrete examples that help you focus on "what is really going on." The other is the formal proof and its associated precision. Both have a role to play, but it is easy to forget about the former when struggling to understand the latter.

2.3 SHIFTING AND DIVISIBILITY

Perhaps we can generate Latin squares by other systematic methods. For example, instead of cyclically shifting by one in each row after the first, what would happen if we tried other shifts? Shifting by two is the obvious next thing to try, so let us begin with that.

Shifting by two works well for Latin squares of orders 3 and 5:

Latin squares of orders 3 and 5 constructed by shifting each row two cells left from the previous row

1	2	3
3	1	2
2	3	1

1	2	3	4	5
3	4	5	1	2
5	1	2	3	4
2	3	4	5	1
4	5	1	2	3

In fact, close inspection reveals that these Latin squares are the same as the one-shift Latin squares we constructed earlier, but with the rows in a different order.

Shifting by two does not work for orders 4 and 6, alas. In these cases, we end up with repeated rows, and thus do *not* obtain Latin squares:

For orders 4 and 6, shifting each row two cells left from the previous row fails to make a Latin square

1	2	3	4
3	4	1	2
1	2	3	4
3	4	1	2

1	2	3	4	5	6
3	4	5	6	1	2
5	6	1	2	3	4
1	2	3	4	5	6
3	4	5	6	1	2
5	6	1	2	3	4

Why the difference? It is because 3 and 5 are odd, while 4 and 6 are even. Shifting by two in a square of even order results in repeated rows, implying that the columns will have repeated entries.

Generalizing from this, if n is a given order, and d is an integer that divides n, then shifting by d will result in repeated rows. Specifically, if $n = ds$, then shifting by d a total of s times will return us to the original row. We just saw this with $n = 6$, $d = 2$, and $s = 3$. Shifting by two, a total of three times, returned us to the original row.

Still more precisely, if k is any integer between 1 and $\frac{n}{d}$ inclusive, then row k will be identical with row $\frac{n}{d} + k$. For example, if $n = 6$ and $d = 2$, then row 1 and row 4 will be identical, as will rows 2 and 5, and rows 3 and 6. You can verify that with the diagram above. Likewise, if we took an order $n = 12$ square constructed by a shift of $d = 3$, then rows 1, 4, 7, and 10 would be identical, as would rows 2, 5, 8, 11, and 3, 6, 9, 12. You should try this yourself to ensure that you agree.

We have established that if d is a divisor of n, then we cannot construct a Latin square of order n by cyclically shifting by d. Does this mean that if d is *not* a divisor of n, then we *can* construct a Latin square of order n with cyclic d-shifts?

Not necessarily. Four does not divide 6, but a shift by four in a square of order 6 leads to this:

Four-shifting in order 6 does not make a Latin square

1	2	3	4	5	6
5	6	1	2	3	4
3	4	5	6	1	2
1	2	3	4	5	6
5	6	1	2	3	4
3	4	5	6	1	2

Note that the fourth row is the same as the first. This is because we arrive at the fourth row above by four-shifting the first row three times. This has the same net effect as shifting by twelve. Since we have six symbols, shifting by twelve – a multiple of six – brings us back to where we started.

Do you see the pattern? If the shift has a divisor (other than 1) in common with the order then we will not obtain a Latin square. Otherwise, we will.

Two positive integers n and d that have no common divisors other than 1 are said to be *relatively prime*. Put differently, the integers n and d are relatively prime if the fraction $\frac{d}{n}$ cannot be reduced. The numbers 4 and 6 are not relatively prime (they have 2 as a common divisor), and that is why we do not obtain a Latin square when we start with six letters and shift by four.

We ought to prove that our pattern holds generally. That is, we want to show that a shift by d in a square of order n produces a Latin square precisely when d and n are relatively prime. Toward that end, let us introduce some new notation. We will denote by

$$\gcd(d, n)$$

the greatest common divisor of the integers d and n. That is, $\gcd(d, n)$ is the largest number that evenly divides both d and n. Saying that d and n are relatively prime is now equivalent to saying that $\gcd(d, n) = 1$. The theorem that we wish to prove is:

Theorem 2

Let n and d be positive integers. Let L be the $n \times n$ matrix whose kth row, for $1 \leq k \leq n$, is obtained by d-shifting the first row to the right a total of $(k-1)$ times. That is, the kth row reads from left to right as

$$[(k-1)d + 1, (k-1)d + 2, \ldots, n, 1, 2, \ldots, (k-1)d].$$

Then L is a Latin square if and only if $\gcd(d, n) = 1$.

We should mention that what follows is a bit more technical than the material to this point. If you find yourself getting bogged down, you can simply skim over it without losing the main thread of the discussion.

Although we have just made our notion of d-shifting more precise, we have also made things more complex. To get a handle on things, notice that Theorem 2 is a generalization of Theorem 1. Equivalently, we could say that Theorem 1 is a special case of Theorem 2; to be precise, it is the special case when $d = 1$.

Let's verify that when $d = 1$, Theorem 2 is exactly the statement of theorem 1. If $d = 1$, then the kth row mentioned in Theorem 2 is:

$$[(k-1)(1)+1, (k-1)(1)+2, \ldots, n, 1, 2, \ldots (k-1)(1)],$$

which easily simplifies to the row

$$(k, k+1, \ldots, n, 1, 2, \ldots, k-1).$$

Moreover, when $d = 1$, for every value of n we have

$$\gcd(d, n) = \gcd(1, n) = 1,$$

since the largest number dividing both 1 and any other positive integer n is the number 1 itself. The conclusion is that when $d = 1$, the matrix with kth row as given above is a Latin square; this is exactly the statement of theorem 1.

We need one more concept before proceeding to our proof of the more general theorem 2. We will need to keep track of what happens to individual entries as we shift them multiple times. For example, suppose we are shifting by four in a square of order 10. After one shift, the entry in column 1 moves to column 4. After two shifts, it will be in column 8. What happens after the third shift?

We would like to say it is in column 12, but there is no such thing. In reality, after passing column 10, the entry cycles back to the beginning, eventually ending in column 2. Notice, though, that 12 is two more than a multiple of 10. It as is if we only kept that part of 12 which is greater than 10. Equivalently, we kept the remainder when 12 is divided by 10.

To discuss this more generally, define the expression

$$A \equiv B \pmod{n}.$$

This is read, "the integer A is congruent to B modulo n." This is a shorthand way of saying that the difference $A - B$ is a multiple of n. For example, we have $21 \equiv 3 \pmod{6}$, since $21 - 3 = 18$ is a multiple of 6.

You are probably familiar with the word *congruent*, from geometry, where congruent triangles, for example, refer to triangles with identical side lengths. The word *modulo*, comes from the Latin word *modulus*, which refers to a standard or measure against which things are compared. In saying that two numbers are "congruent modulo n," we are saying they are identical when compared using n as the standard.

Let us try an exercise to test our understanding. Suppose n and d are two positive integers. Can we find distinct positive integers A and B between 0 and $n - 1$ inclusive with the property that

$$Ad \equiv Bd \pmod{n}?$$

Some experiments are in order. If $d = 4$ and $n = 6$, then our congruence becomes

$$4A \equiv 4B \pmod{6}.$$

In this case, we can see that $A = 3$ and $B = 0$ will get the job done, since $12 \equiv 0 \pmod{6}$.

On the other hand, if $d = 5$ and $n = 7$, then our congruence is

$$5A \equiv 5B \pmod{7}.$$

Our search for A and B will be futile in this case. You can convince yourself of that by noting that A and B are only permitted to take on the values 0–6, and then trying them all.

It turns out that such A and B exist precisely when d and n fail to be relatively prime. Notice that our congruence is equivalent to saying that

$$Ad - Bd = (A - B)d$$

is a multiple of n. If d and n have no common divisor other than 1, then $A - B$ must be a multiple of n. But since A and B are both smaller than n, this is possible only if $A = B$ (in which case, $A - B = 0$). This shows there is no solution when d and n are relatively prime.

Now suppose that d and n are not relatively prime. In this case, there is a number m other than 1 that divides them both. We can then let A and B be the integers

$$A = \tfrac{n}{m} \quad \text{and} \quad B = 0.$$

This is the formula we used to generate the solution to the previous example. We have

$$Ad - Bd = (\tfrac{n}{m})d - 0d = (\tfrac{n}{m})d,$$

which is an integer multiple of d, since $A = \frac{n}{m}$ is an integer.

We are now ready to prove Theorem 2, which if you recall, claimed that our d-shifting technique for constructing a square of order n results in a Latin square precisely when $\gcd(d, n) = 1$.

Proof

Again we must show that each of the numbers $1, 2, \ldots, n$ appears exactly once in each row and column. For rows this is true regardless of the value of d, since by construction each row consists of the numbers from $(k-1)d+1$ to n followed by the numbers from 1 to $(k-1)d$, for some value of k.

Now let us consider the first column of our constructed matrix L. According to our procedure, the entries in this column are

$$[1, (2-1)d+1, (3-1)d+1, \ldots, (n-1)d+1],$$

where each entry is considered modulo n. Simplifying gives us

$$[0d+1, 1d+1, 2d+1, 3d+1, \ldots, (n-1)d+1],$$

From this we see that column 1 will have repeated entries when there are two distinct integers A and B, between 0 and $n-1$ inclusive, for which

$$Ad+1 \equiv Bd+1 \pmod{n}.$$

This is equivalent to saying that $Ad \equiv Bd \pmod{n}$, which is possible only if d and n are relatively prime.

We have now established that the first column of our d-shift matrix contains each of the numbers $1, 2, \ldots, n$ exactly once precisely when d and n are relatively prime. Since each subsequent column is a shift of the first column, the proof is complete. □

2.4 JUMPING IN THE RIVER

There are two things to notice about this discussion. First, we did not introduce any mathematical symbols or notation until we needed them. Symbols are just abbreviations, and they are used in mathematical writing solely to improve the economy of the presentation. If you attempt to rewrite our argument without the benefit of mathematical notation, you will quickly come to appreciate its value.

Unfortunately, mathematical symbols can seem like hieroglyphics until you have fully internalized them. This can present quite a challenge to students of our discipline. At the same time you are trying to think clearly about unfamiliar, abstract problems, you are also trying to master the foreign language in which mathematical texts are written. Were there a simple resolution to this dilemma, math teachers would have employed it long ago. We can only comfort you with the thought that with practice, the symbols come to seem natural.

Second, notice the flow of our discussion. In considering our elementary questions about Latin squares, we were led unavoidably to questions about congruences and greatest common divisors. These are topics typically discussed in courses on elementary number theory. It had not been our specific intention to discuss number theory, but it was essential to solving our problem. This sort of thing happens all the time in mathematical research. The goal is to solve the problem, and toward that end you will make use of whatever tools are appropriate to the job.

The whole thing is like a river. The problem resides at the river's source, and this is where you jump in. From there you are largely at the mercy of the current, going wherever it takes you. Often you end up far from any destination you envisioned when jumping in. But as with other sorts of travel, the journey is half the fun.

3

Greco-Latin Squares

The Problem of the Thirty-Six Officers

To this point, our squares have been constructed from a single set of symbols, namely the numbers from 1 to n. What if we considered two sets of symbols? Could we arrange things so that in addition to having a Latin square in each set of symbols individually, each possible pair of symbols appears only once? Our next puzzle will clarify our intent.

Puzzle 17: Royalty Sudoku.

Suppose we remove from a standard fifty-two-card deck the twelve face cards and the four aces; that is the ace, king, queen, and jack of each of the four suits. Arrange these cards in a 4×4 square so that each denomination and each suit appears exactly once in every row and column. One card has been placed in the grid to get you started. A solution is in the text below.

A♠			

3.1 DO GRECO-LATIN SQUARES EXIST?

Prior to revealing the solution to Puzzle 17, let us consider what happens in the slightly simpler 3×3 case. Suppose we want to fill a 3×3 grid so that each cell has two symbols, one from the set A, B, C and one from the set α, β, γ. Further, suppose we wish to place each of the possible $(3)(3) = 9$ ordered pairs consisting of a Latin letter A, B, or C followed by a Greek letter α, β, or γ exactly once in the grid. For example, the ordered pair $A\alpha$ should appear once and only once. There are many possible ways of doing this, one of which is the following:

$A\alpha$	$B\beta$	$C\gamma$
$B\gamma$	$C\alpha$	$A\beta$
$C\beta$	$A\gamma$	$B\alpha$

Notice that the figure is made of two superimposed Latin squares, one involving A, B, C and one involving α, β, γ:

A	B	C
B	C	A
C	A	B

α	β	γ
γ	α	β
β	γ	α

When the two squares are superimposed, we obtain each of the nine possible ordered pairs. Pairs of squares with this property are said to be *orthogonal*. That is, two Latin squares of order n are said to be orthogonal if the square obtained by superimposing them contains each possible ordered pair exactly once. Leonhard Euler introduced the study of orthogonal Latin squares in a paper published in 1782. Just as we did above, he used Greek letters for one of the squares and Latin letters for the other. For that reason, the object obtained by superimposing the two squares is today referred to as a *Greco-Latin square*.

Now let us return to the Royalty Sudoku puzzle. Given the relatively small number of cards, a little judicious experimentation ought to produce an answer. There are many possible solutions, one of which is the following:

One way of completing the Royalty Sudoku puzzle

A♠	K♡	Q♢	J♣
Q♣	J♢	A♡	K♠
J♡	Q♠	K♣	A♢
K♢	A♣	J♠	Q♡

Once again, notice that we have superimposed two Latin squares, one involving the denominations and one involving the suits:

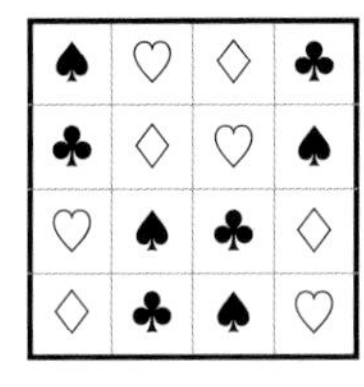

Following the lead of the previous section, we might now wonder whether Greco-Latin squares of arbitrary order exist. We have seen they exist for orders 3 and 4, but what about other orders? For example, what about an order 5 Greco-Latin square? In the following puzzle, you will construct just such an object. To make the puzzle easier to work with we use letters and numbers instead of Latin and Greek letters.

Puzzle 18: Greco-Latin-Doku.

Fill in the grid so that in the upper-left white corners, every row and column contains *A–E* exactly once, and in the lower-right green corners, every row and column contains 1–5 exactly once. In addition, each letter-number combination must appear exactly once on the board. (So for example, one and only one cell can have C in the upper-left with a 2 in the lower-right.)

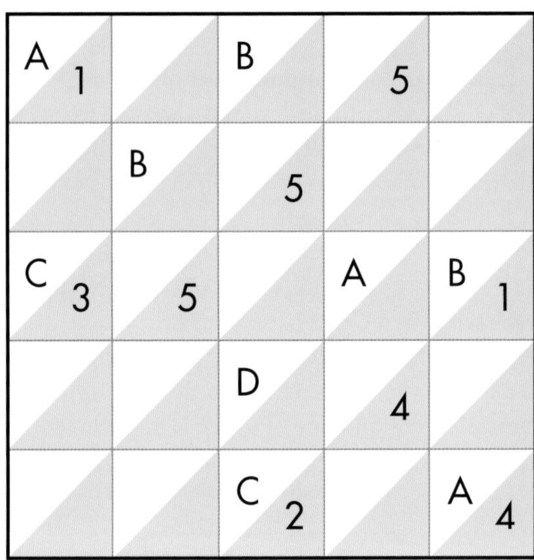

Things are looking pretty good at this point. We have already established that Greco-Latin squares – or equivalently, pairs of orthogonal Latin squares – exist of orders 3, 4, and 5. But are there such squares for *all* orders?

The answer is no, and we do not have to look hard to find a counterexample. There is no pair of orthogonal 2 × 2 Latin squares. With so small an order, our options are limited to the point that we can simply list all the possibilities. The individual squares must look essentially like this:

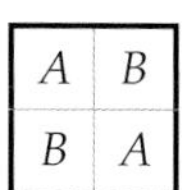

α	β
β	α

We could use different symbols, of course, perhaps C and D in place of A and B, or γ and δ in place of α and β. Still, whatever symbols are used, a 2×2 Latin square must have one of the symbols along one of the diagonals and the other symbol along the other. When they are superimposed, we obtain a square of the form shown below.

Ack, repeated pairs!

$A\alpha$	$B\beta$
$B\beta$	$A\alpha$

Since we now have repeated ordered pairs, the square above is not a Greco-Latin square. Therefore, there is no Greco-Latin square of order 2.

Perhaps the 2×2 case is exceptional, given its dimunitive size? Sadly, the situation is worse than that. It turns out that there is no Greco-Latin square of order 6. There is irony in this, since the problem of constructing such a square was the motivation for Euler's work, as we shall discuss shortly.

Now that we know we cannot expect to produce Greco-Latin squares for arbitrary orders, we can change our question somewhat. We now ask: For which values of n are there Greco-Latin squares of order n? When $n = 3, 4, 5$ they exist, while for $n = 2, 6$ they do not. That leaves rather a lot of territory currently unexplored. To help chip away at things, perhaps we can find a procedure that generates squares of certain sizes, even if there is no hope of finding a procedure that works for arbitrary values of n.

Let us start with the case where n is odd. In this case, we can follow a definite procedure for constructing an order n Greco-Latin square. Consider the pattern in the following order 3 and order 5 Greco-Latin squares:

Shifting the Latin parts by one and the Greek parts by two

$A\alpha$	$B\beta$	$C\gamma$
$B\gamma$	$C\alpha$	$A\beta$
$C\beta$	$A\gamma$	$B\alpha$

$A\alpha$	$B\beta$	$C\gamma$	$D\delta$	$E\varepsilon$
$B\gamma$	$C\delta$	$D\varepsilon$	$E\alpha$	$A\beta$
$C\varepsilon$	$D\alpha$	$E\beta$	$A\gamma$	$B\delta$
$D\beta$	$E\gamma$	$A\delta$	$B\varepsilon$	$C\alpha$
$E\delta$	$A\varepsilon$	$B\alpha$	$C\beta$	$D\gamma$

The Latin parts of the squares above are constructed by the "shift by one" method of the previous section. In the Greek parts of each square, each row is shifted by two from the one before it. We know from our earlier examination of greatest common divisors that shifting by two produces a Latin square only when n is odd, and that is why this method of constructing Greco-Latin squares can only work for odd orders.

When we superimpose squares constructed in this way, is the result always a Greco-Latin square? It certainly worked for $n = 3$ and $n = 5$. For $n = 7$, the following square results:

Doing the Latin one-shift with the Greek two-shift in order 7

$A\alpha$	$B\beta$	$C\gamma$	$D\delta$	$E\varepsilon$	$F\zeta$	$G\eta$
$B\gamma$	$C\delta$	$D\varepsilon$	$E\zeta$	$F\eta$	$G\alpha$	$A\beta$
$C\varepsilon$	$D\zeta$	$E\eta$	$F\alpha$	$G\beta$	$A\gamma$	$B\delta$
$D\eta$	$E\alpha$	$F\beta$	$G\gamma$	$A\delta$	$B\varepsilon$	$C\zeta$
$E\beta$	$F\gamma$	$G\delta$	$A\varepsilon$	$B\zeta$	$C\eta$	$D\alpha$
$F\delta$	$G\varepsilon$	$A\zeta$	$B\eta$	$C\alpha$	$D\beta$	$E\gamma$
$G\zeta$	$A\eta$	$B\alpha$	$C\beta$	$D\gamma$	$E\delta$	$F\varepsilon$

More success! Indeed, if you study these examples carefully, it becomes clear that this method will work for any odd value of n. To establish this, we need only make sure that each pairing of a Latin letter and a Greek letter occurs exactly once. Notice, though, that the Latin letters get shifted by one, while the Greek letters get shifted by two, as compared to the row above it. This is effectively equivalent to leaving the Latin letters fixed, while the Greek letters get shifted by one beside them. Viewed in this way, it is clear that we do not get any repeated pairs.

The solution for even values of n proves to be far more difficult, and it would take us too far afield to investigate that problem here. Let us instead consider the history of this problem, to see what it tells us about mathematics.

3.2 EULER'S GRECO-LATIN SQUARE CONJECTURE

The following account is drawn largely from the article by Klyve and Stemkoski [27] and the book by Stinson [39].

Euler introduced his 1782 paper on this topic [22] as follows:

> *A very curious question that has taxed the brains of many inspired me to undertake the following research that has seemed to open a new path in analysis and in particular in the area of combinatorics. This question concerns a group of thirty-six*

officers of six different ranks, taken from six different regiments, and arranged in a square in a way such that in each row and column there are six officers, each of a different rank and regiment.

In our terminology, he sought a Greco-Latin square of order 6, where rank and regiment are represented by the Greek and Latin letters, respectively. Euler gives no indication of the source for the problem. Among the many accomplishments in the paper, Euler established that Greco-Latin squares exist for all odd values of n, and also for all values of n that are multiples of 4. He then conjectured that Greco-Latin squares did *not* exist for orders such as 6, 10, and 14 which leave a remainder of 2 when divided by 4.

Little progress was made for over a century, until the French mathematician Gaston Tarry [44, 45] established the nonexistence of Greco-Latin squares of order 6. Tarry's work was published in 1900 and 1901. His work was a masterpiece of persistence. It involved an exhaustive list of more than 9, 000 possibilities.

It was not until 1984 that Stinson produced a theoretical proof of this result [40]. Another theoretical proof, using entirely different techniques, was published by Dougherty in 1994 [20]. Which raises an important question: What was the point of publishing new proofs of a result that was already known? Tarry established in 1901 that there was no Greco-Latin square of order 6. Why was that not the end of the story?

The answer is that a good proof has value beyond establishing the truth of a proposition. It is one thing to know *that* something is true, but it is quite another to understand *why* it is true. It is useful to know a car will move when you step on the gas, but it is even more useful to understand why stepping on the gas causes the car to move. Each proof tells us something we did not know before: not about configurations of soldiers, but about the usefulness of the techniques that are used.

What of the rest of Euler's conjecture? He had been proven correct about nonexistence in the case $n = 6$, but what about the remaining cases such as $n = 10, 14, 18$, and so on? The next big step came in a 1922 paper by MacNeish [28]. He devised a method for constructing large Greco-Latin squares from small ones, just as you might construct a large wall by stacking up small bricks. His construction depended on the idea of the direct product $S \times T$ of two Latin squares S and T. To see how it works, consider the following example:

S

A	B	C
B	C	A
C	B	A

T

α	β
β	α

$S \times T$

$A\alpha$	$A\beta$	$B\alpha$	$B\beta$	$C\alpha$	$C\beta$
$A\beta$	$A\alpha$	$B\beta$	$B\alpha$	$C\beta$	$C\alpha$
$B\alpha$	$B\beta$	$C\alpha$	$C\beta$	$A\alpha$	$A\beta$
$B\beta$	$B\alpha$	$C\beta$	$C\alpha$	$A\beta$	$A\alpha$
$C\alpha$	$C\beta$	$A\alpha$	$A\beta$	$B\alpha$	$B\beta$
$C\beta$	$C\alpha$	$A\beta$	$A\alpha$	$B\beta$	$B\alpha$

It is as if each entry of S gets its own copy of T attached to it (mentally divide $S \times T$ above into 2×2 blocks and you will see what we mean). The result is a Latin square of order 6 whose elements are pairs. (Not *ordered* pairs, just pairs of symbols that together represent one character; notice, for example, that the element $A\alpha$ appears exactly once in each row and column.)

MacNeish showed that if S_1 and S_2 are a pair of orthogonal Latin squares, and if T_1 and T_2 are a second pair of orthogonal Latin squares, then the square $S_1 \times T_1$ is orthogonal to $S_2 \times T_2$. Using his technique, we could, for example, take our earlier examples of order 4 orthogonal Latin squares and order 7 orthogonal Latin squares, and use them to manufacture a pair of orthogonal Latin squares of order 28.

Progress continued to be made throughout the 1930s and 1940s, as mathematicians attacked the problem with ever more sophisticated techniques. The punch line came in 1960, when Bose, Parker, and Shrikhande [14] resolved the problem once and for all. Building on the work that came before them, they established that Greco-Latin squares exist for *all* values of n except for 2 and 6. Euler's conjecture was false. This result was considered so significant that it was featured on the front page of *The New York Times*.

That's what mathematics is really all about. As a professional mathematician your job is to solve problems. You are not working in a vacuum, however. Presumably others are interested in the same problems as you, and it is part of your job to be up on the latest developments. Gradually you chip away at a problem until so little remains unknown that it is ready to fall altogether. Solutions to big problems are nearly always the end results of years of incremental progress, with contributions from dozens of mathematicians.

There is a large component of solitude in mathematical work. There is also, however, a significant communal aspect. You are part of a community of scholars all working toward a common end. There is great satisfaction in this, even if at times the work can seem esoteric and far removed from everyday concerns.

Unfortunately, this is rarely seen or experienced by those outside the mathematical community. After years of classes in which mathematics is presented as nothing more than symbol manipulation and mindless rule-following, it is small wonder most people are puzzled that anyone does such things voluntarily. This distaste is far more common among those who have never experienced the real thing.

3.3 MUTUALLY ORTHOGONAL GERECHTE DESIGNS

With Euler's conjecture resolved, mathematicians proceeded to the logical next step. Orthogonal Latin squares exist for all orders except 2 and 6. Very nice – but let's take it one step further. For example, for a given n, how many pairwise orthogonal Latin squares can you have at the same time? A set of Latin squares is said to be *mutually orthogonal* if any two of them are orthogonal to each other. As an example, the following set of four 5×5 Latin squares is mutually orthogonal. If you compare any two of the squares you will find that they are orthogonal.

Any two of these grids would contain each possible ordered pair exactly once if superimposed

A	B	C	D	E
B	C	D	E	A
C	D	E	A	B
D	E	A	B	C
E	A	B	C	D

A	B	C	D	E
C	D	E	A	B
E	A	B	C	D
B	C	D	E	A
D	E	A	B	C

A	B	C	D	E
D	E	A	B	C
B	C	D	E	A
E	A	B	C	D
C	D	E	A	B

A	B	C	D	E
E	A	B	C	D
D	E	A	B	C
C	D	E	A	B
B	C	D	E	A

Is it possible to find a fifth 5×5 Latin square that is orthogonal to each of the other four? It is not, but proving that fact is more trouble than it is worth here. The proof follows from a more general result asserting that for any value of $n \geq 2$, it is impossible to have more than $n - 1$ mutually orthogonal Latin squares. There might be fewer than $n - 1$, but there are definitely no more than that.

For example, there can only be up to four mutually orthogonal 5×5 Latin squares, and only up to eight mutually orthogonal 9×9 Latin squares. It is also known that values of n that are either prime or powers of a prime number attain that upper bound. For example, since 5 is prime, we were able to find four mutually orthogonal 5×5 Latin squares. And since $9 = 3^2$ is a power of a prime, there exists a set of eight mutually orthogonal 9×9 Latin squares. Proofs of these assertions can be found in Stinson [39].

For other sorts of values of n we cannot be certain that the upper bound of $n - 1$ mutually orthogonal Latin squares is attained. For example, there could be as many as five 6×6 Latin squares, or there could be fewer than that.

Now, what if we restricted our attention to Latin squares with additional block conditions? Do such extra conditions affect our ability to produce orthogonal squares?

Indeed it does. In the $n = 5$ case, we could consider a 5×5 Latin square subdivided into regions as shown in Puzzle 19. If we require the numbers 1–5 to appear exactly once in each row, column, and region, then we have what we will call a *Crossdoku* square.

Puzzle 19: Crossdoku.

Fill in each grid so that every row, column, and marked region contains the numbers 1–5 exactly once.

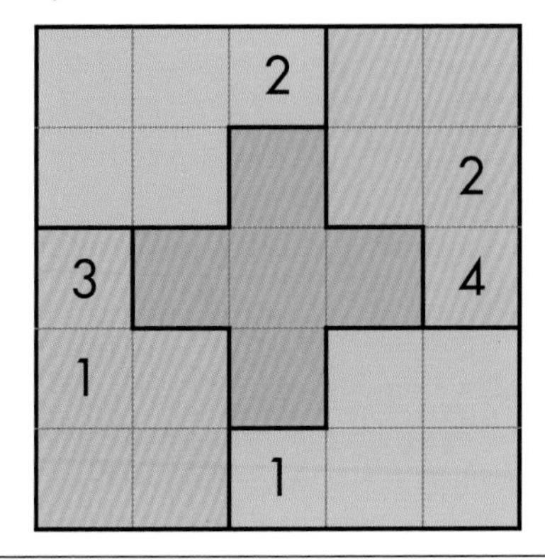

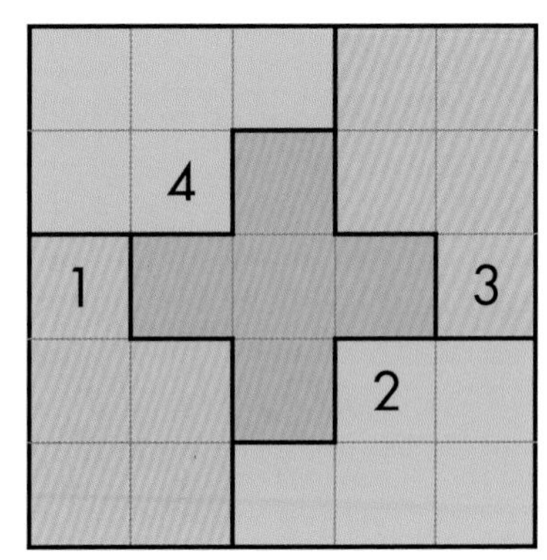

How many mutually orthogonal Crossdoku boards could we have? We can answer this with a more general theorem. First, a definition. An $n \times n$ *Gerechte design* is an $n \times n$ Latin square that has been further subdivided into n additional regions of size n. In an $n \times n$ Gerechte design, the numbers 1–n must each appear exactly once in every row, column, and additional region. Sudoku squares are examples of 9×9 Gerechte designs. The Crossdoku squares above are 5×5 Gerechte designs.

In a recent paper by Bailey, Cameron, and Connelly [7] it was shown that given a particular type of $n \times n$ Gerechte design, the number of mutually orthogonal squares is bounded above by $n - d$, where d is the largest overlap between one of the additional regions and some row or column in the square. For our Crossdoku puzzle, we have $d = 3$, since several of the regions have three cells in common with various rows and columns. (For example, the central cross-shaped region has three cells in common with the third row of the grid). Therefore the maximum number of mutually orthogonal Crossdoku boards that we could ever hope for would be

$$n - d = 5 - 3 = 2.$$

That is the theoretical maximum, but our theorem does not guarantee this maximum is actually attained. In this case, the answer is less than the maximum. It happens that there are *no* mutually orthogonal Crossdoku boards. The reason for this involves the following puzzle:

Puzzle 20: Related Crossdoku Cells.

Consider the nearly empty Crossdoku grid below. If the entry in the third cell in the second row is X, then what can you say about the entry in the last cell in the third row, and why?

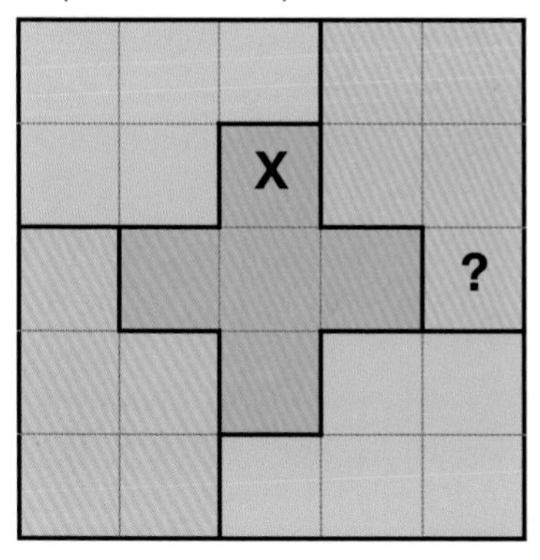

In the puzzle above, you will find that there must be an X in the cell with the question mark. So if there were ever a pair of orthogonal Crossdoku boards, when superimposed we would have some pair (X, Y) in the third cell of the second row, and again the same pair (X, Y) in the last cell of the third row. Since orthogonal boards cannot have repeated ordered pairs, we cannot have a pair of orthogonal Crossdoku boards.

We are more fortunate in the case of order $n = 4$ Greco-Latin squares. As you shall see in the next puzzle, there do exist Greco-Latin 4×4 Gerechte designs with 2×2 blocks. Since the largest overlap between the blocks and the rows and columns is $d = 2$, we have achieved the maximum possible number $n - d = 2$ of mutually orthogonal designs of this type.

Puzzle 21: Greco-Latin Mini-Sudoku.

Fill in each grid so that every row, column, and block contains A–D exactly once and 1–4 exactly once. In addition, each letter-number combination must appear exactly one time on the board.

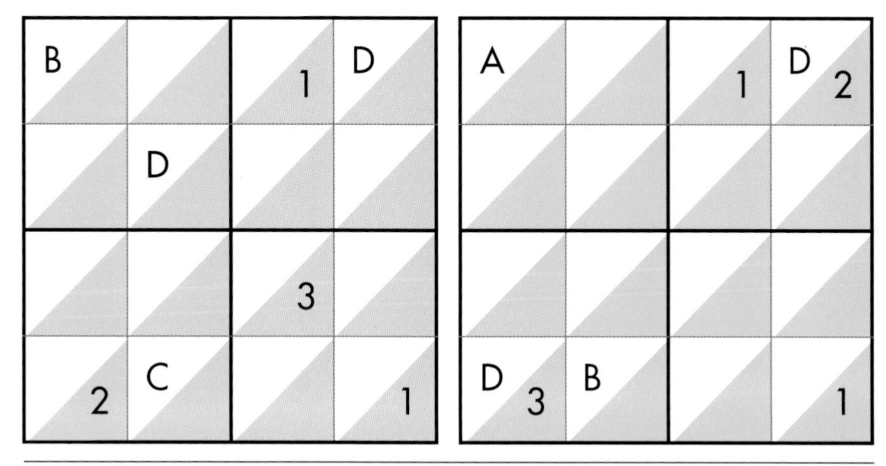

3.4 MUTUALLY ORTHOGONAL SUDOKU SQUARES

What happens in the 9×9 Sudoku case? Let us start our investigation with a 9×9 Sudoku variation based on a pair of mutually orthogonal Sudoku squares. Taken separately, neither the white Sudoku square nor the green Sudoku square has a unique solution. However, when considered together, they make a pair of orthogonal squares. The extra orthogonality condition ensures a uniquely correct way of filling in both grids simultaneously.

Puzzle 22: Greco-Latin Sudoku.

Fill in the grid so that every row, column, and block contains *A–I* exactly once and 1–9 exactly once. In addition, each letter-number combination must appear exactly one time on the board.

	2		8	B 3			6	
G 6		1	E		A 7			F 8
9		F				E 5		2
	D		A 1		B 5		G	
	G 5			F 4			E 8	
	7		D 3		G 8		4	
F		G 9				4		C
D 2			C 5		4	B		G 7
	I			G 2	H		D	

How many mutually orthogonal Sudoku squares could we have? We already know there are eight mutually orthogonal Latin squares of order 9 (which are Sudoku squares without the block condition.) However, the 3×3 blocks have an overlap of three cells with any row or column. If we apply our previous theorem with $d = 3$, then the maximum number is no more than $9 - 3 = 6$.

Happily, this maximum is attainable. An example of six mutually orthogonal Sudoku squares is shown in the following diagram, which is a modification of an example from Bailey, Cameron, and Connelly [7]. All six squares are shown here superimposed onto one 9×9 grid. We have added a color coding for each number 1–9 so that the patterns are more readily visible. Each cell in the big grid has six compartments, each representing a cell of one of the Sudoku squares. For example, the set of upper-left numbers make a Sudoku square, as do the set of center numbers, the set of lower-right numbers, and so on. In this figure, the central cells all have the same number in each of their compartments. That does not have to be the case, but was done in this example for visual appeal.

Six mutually orthogonal Sudoku squares, superimposed and shown with the nine numbers represented by nine colors

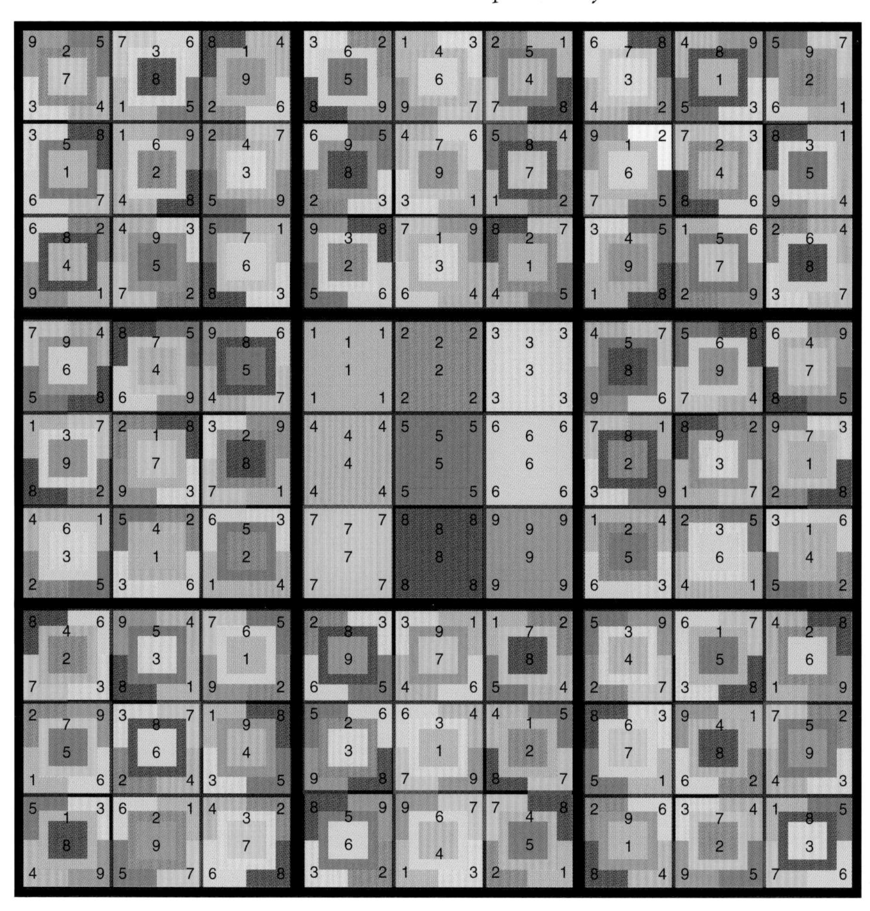

3.5 WHO CARES?

To this point we have mostly been free-forming, asking questions about Latin squares and related objects with no motivation beyond our own amusement.

In doing so, we were following Euler's example. He closed his paper by writing:

> *Here I bring mine to an end on a question that, although it is of little use itself, has led us to some observations as important for the doctrine of combinatorics as for the general theory of magic squares.*

Still, while the value of intellectual amusement can hardly be minimized, it is also nice to ponder things of practical importance. As it happens, Latin and Greco-Latin squares find a variety of applications. We will now describe two of them.

Latin squares have been used in the statistical design of experiments going back at least to the 1930s. To illustrate the idea, suppose we have five varieties of gasoline that we want to test for their efficiency. (This example is taken from the book by Bogart [12].) Since weather conditions change from day to day in ways that can affect fuel efficiency, we shall conduct our test using five different cars driving a fixed route on the same day.

This introduces a new problem: Different cars have different fuel mileage characteristics. To remedy this, we will perform the experiment five times, making sure that each fuel is placed in each car exactly once. A Latin square of order 5 can then be interpreted as a schedule for which gasoline goes in which car on which day. Let us refer to this square as S for schedule. One possible schedule S could look like this:

	Mon	**Tues**	**Wed**	**Thurs**	**Fri**
Chevy	gas 1	gas 2	gas 3	gas 4	gas 5
Ford	gas 2	gas 3	gas 4	gas 5	gas 1
Honda	gas 3	gas 4	gas 5	gas 1	gas 2
Toyota	gas 4	gas 5	gas 1	gas 2	gas 3
Volvo	gas 5	gas 1	gas 2	gas 3	gas 4

Suppose we are also concerned that fuel efficiency can vary depending on who is driving the car. We want each of our five drivers to encounter each variety of gasoline exactly once, and to drive each car exactly once. A square orthogonal to S would then provide a scheme for ensuring that our requirements are met:

	Mon	Tues	Wed	Thurs	Fri
Chevy	gas 1 Adam	gas 2 Beth	gas 3 Carl	gas 4 Dave	gas 5 Erin
Ford	gas 2 Carl	gas 3 Dave	gas 4 Erin	gas 5 Adam	gas 1 Beth
Honda	gas 3 Erin	gas 4 Adam	gas 5 Beth	gas 1 Carl	gas 2 Dave
Toyota	gas 4 Beth	gas 5 Carl	gas 1 Dave	gas 2 Erin	gas 3 Adam
Volvo	gas 5 Dave	gas 1 Erin	gas 2 Adam	gas 3 Beth	gas 4 Carl

This can be pushed further, of course. Suppose we also want our cars tested on five different kinds of road. Then a third Latin square, orthogonal to the first two, would provide a schedule. Larger sets of mutually orthogonal Latin squares would be useful for more complex experiments.

Not bad, but how about a more modern example?

Consider the problem of communicating data across power lines. When you use the Internet, data is transmitted between computers through some sort of cable. Fiber optic cables are generally used for this purpose. Electrical impulses are converted by a transmitter into pulses of light. These pulses travel along the cable to the receiving computer, where they are converted back into electrical impulses.

This all works very well, but it requires laying down huge numbers of fiber optic cables. It also implies certain practical limits on the quantity of information that can be transmitted. This quantity of information, known as bandwidth, is limited by the amount of cable you have. Since demand for bandwidth is always likely to outstrip supply, it makes sense to look for other sorts of communications channels.

An obvious candidate is the elaborate network of copper and coaxial cables that already transmit electricity to our homes. Why not employ this already existing electrical infrastructure for transmitting data?

The problem is noise in the channel. Electrical impulses sent along traditional copper or coaxial cables get distorted by a wide variety of natural phenomena. By the time they arrive at your computer, the signal is too corrupted to be translated back into meaningful data. You are familiar with the basic phenomenon from having heard static on the radio. This is not a problem for bringing power to your home, but it poses a real challenge for data transmission. Fiber optic cables are far less susceptible to this problem.

A possible solution involves encoding the data in such a way that the receiver can still interpret the signal despite the noise pollution. Since existing codes are insufficient for dealing with the many sources of noise to which traditional cables are subject, more sophisticated techniques are needed. The most promising solution involves mutually orthogonal Latin squares.

Since different sources of noise corrupt signals at different frequencies, it makes sense to use many different frequencies for sending our signals. Noise might corrupt some of the frequencies, but we can hope that enough remain uncorrupted to preserve the signal's meaning. This strategy is borrowed from radio where it is referred to as "frequency modulation," usually abbreviated as FM. As an analogy, think about reading a book. A handful of typographical errors, while annoying, do not render the text unreadable. But if there are many typographical errors, or if whole sequences of words are placed out of order, then the message is quickly lost.

We now have conflicting concerns. More frequencies permit more information to be sent, but at the cost of consuming more bandwidth. We need codes that balance these concerns, and that is where mutually orthogonal Latin squares come in.

Let us imagine we have four frequencies with which to work, called F1, F2, F3, and F4. On each frequency we can send a pulse. To encode a letter of the alphabet we can send four pulses, one on each frequency, in some order. For example, we could encode the first twelve letters of the alphabet as follows:

Frequency codes for the first twelve letters of the alphabet

	F1	F2	F3	F4
A	1	2	3	4
B	2	1	4	3
C	3	4	1	2
D	4	3	2	1

	F1	F2	F3	F4
E	1	3	4	2
F	2	4	3	1
G	3	1	2	4
H	4	2	1	3

	F1	F2	F3	F4
I	1	4	2	3
J	2	3	1	4
K	3	2	4	1
L	4	1	3	2

Under this system the letter *A* would be transmitted by sending a pulse at frequency 1, then frequency 2, then 3, and then 4 (note the first row of the first table). To transmit *B*, we send a pulse along frequencies 2, 1, 4, and 3, in that order (see the second row of the first table). And so on. Notice that the three squares above comprise a set of mutually orthogonal Latin squares.

What makes this code so special? It is the fact that it is possible to detect, and correct, up to two errors in each transmitted letter.

Puzzle 23: Spurious Signals.

Suppose you send a transmission to a friend using the code above. You want to send one letter to your friend, but some source of noise causes spurious signals to appear at frequencies 1 and 2. In other words, in each slot of your transmission your friend receives an extra signal of 1 and 2 in addition to the intended signal. Your friend receives the following message from you:

312, 12, 12, 124

Is your friend able to determine which letter you intended to transmit? If so, which letter is it? The solution is in the text below.

In this puzzle, your friend expected a sequence of four individual frequencies, but also received some extra noise. We know that the letters are encoded so that each frequency is used exactly once. To find the intended list of four numbers, think of it like Sudoku – the numbers 1–4 must each appear exactly once, so 1 and 2 must be in the center two slots in some order, with 3 and 4 on the ends. The only row in the tables given above with this pattern is the one for letter G. Thus, your friend still gets the message, even with all the noise in frequencies 1 and 2.

Larger sets of mutually orthogonal Latin squares could encode even more data and be even less sensitive to noise pollution. We invite you to peruse the very readable article by Stewart [38] (from which the above example was taken), and the more detailed article by Huczynska [25] to learn more about this.

Which brings us to our final point about mathematics, at least for this chapter. The application of Latin squares to the design of experiments arose roughly a century and a half after Euler wrote his paper. The use of mutually orthogonal Latin squares as a strategy for coding data transmissions is a major area of research today. Yet Euler surely did not have any of this in mind in undertaking his work.

This is often how it works. Abstract mathematics studied with no practical application in mind is later seen to be just what is needed to solve a real-world problem. This is a source of great satisfaction to every mathematics educator who has ever had to answer an annoyed student asking, "Who cares about any of this?" You undertake mathematical research partly because it is fun and intellectually satisfying, but you also do it because you do not know from where the next great idea is coming. No explorer knows where his work will lead him, but he persists with the confidence that it is useful to know what is out there.

4

Counting

It's Harder than It Looks

Suppose you have a bowl of apples, and you want to determine how many there are. You would likely remove the apples sequentially from the bowl, labeling each with a particular natural number. The first apple would be labeled "one," the second would be labeled "two," and so on. This process is known as counting. It is something we all learned at a young age.

Counting is not always so straightforward. If the set you are counting is very large, then there is little hope of assigning a specific number to each of its elements. Try counting the grains of sand on the local beach and you will see what we mean.

With newspapers and puzzle magazines churning out a steady stream of Sudoku puzzles it would be nice to know if they are going to run out anytime soon. Is there any danger that every possible Sudoku square will eventually appear as the solution to some puzzle in a magazine? Just how many Sudoku squares are there?

The present chapter is devoted to that question. Since the answer is a very large number, we will not be able to find it just by listing all of the possibilities. Something more clever is called for.

4.1 HOW TO COUNT

Since we cannot count Sudoku boards directly, we will need some strategies for counting them indirectly. Two basic techniques in particular will be helpful, and we introduce them here.

The first is this: If you are carrying out a process with k different steps, and there are n_i ways of carrying out the ith step, with $1 \leq i \leq k$, then there are $n_1 n_2 \ldots n_k$ ways of carrying out the whole process. As we have seen previously, straightforward ideas can be made to seem complicated when stated in precise, technical prose, and that is what we have here. We will not provide a formal proof, since an example will make things clear.

Suppose you are at an ice cream parlor at which you can order one, two, or three scoops in a cone, with a choice of either all chocolate scoops or all strawberry scoops, and with or without sprinkles. Then our basic principle tells us there are

$$3 \times 2 \times 2 = 12$$

different cones we could order. Ordering is a three-step process (choose a size, choose a flavor, choose whether or not to have sprinkles), illustrated by the following diagram:

Twelve ice cream possibilities: three sizes, two flavors, two topping choices

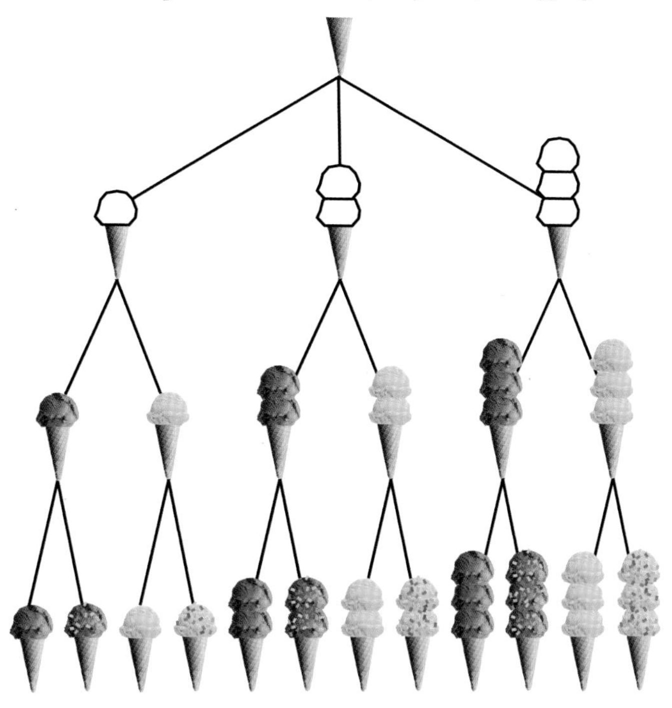

Our second principle is: Given n objects, there are $n!$ ways of arranging them in a straight line. The symbol $n!$, read "n factorial," means to multiply together all of the natural numbers from 1 to n inclusive. Thus, we have $5! = 120$ since

$$5! = 1 \times 2 \times 3 \times 4 \times 5 = 120.$$

The logic here is that any of the n objects can come first in line, then any of the remaining $n - 1$ objects can appear second, and so on. Lining up n objects can be viewed as an n-step process in which there are n ways of carrying out the first step, $n-1$ ways of carrying out the second, and so on. Our first principle now tells us there are $n!$ total ways of lining things up.

Once more, an example will make things clear. According to our principle, there should be $4! = 24$ ways of arranging the letters A, B, C, and D. This is easily verified by listing all of the possibilities. Here they are in alphabetical order:

ABCD	*ABDC*	*ACBD*	*ACDB*	*ADBC*	*ADCB*
BACD	*BADC*	*BCAD*	*BCDA*	*BDAC*	*BDCA*
CABD	*CADB*	*CBAD*	*CBDA*	*CDAB*	*CDBA*
DABC	*DACB*	*DBAC*	*DBCA*	*DCAB*	*DCBA*

We will be putting both of these principles to good use.

If you are not accustomed to the idea that counting can be tricky, we ask you to consider two classic counting problems.

Puzzle 24: Checkerboard Squares.

How many squares are there, of any size, in a 5×5 checkerboard? (Hint: Consider squares of size 1×1, 2×2, 3×3, and so on.)

Now that you've had some practice counting squares, let's move up to a standard 8×8 checkerboard. How many squares, of any size, can we find in the checkerboard below left? We can see at a glance that there are sixty-four small (1×1) squares. But what about squares of other sizes? We must also count the green-tinted 5×5 square shown below right, for example.

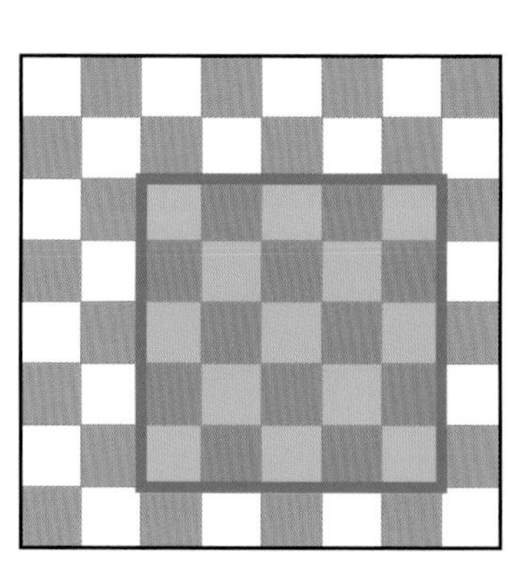

There are so many squares to be counted that, lacking some orderly procedure for listing them all, we will quickly lose track of which we have counted and which we have not. What is needed is some way of breaking the problem into more manageable pieces. For convenience, we shall label the columns from left to right as a through h, while the rows shall be numbered, from bottom to top, as 1 through 8.

What if we restrict our attention to squares of a particular size? We have already seen there are sixty-four small, 1×1 squares. How many 2×2 squares are there?

To answer that we might notice that every 2×2 square has exactly one lower-left corner, and any 1×1 square can be the lower-left corner of only one 2×2 square. It follows that we can determine the number of 2×2 squares by counting the number of 1×1 squares that can serve as a lower-left corner. For example, the green-tinted square below can be uniquely associated with its lower-left corner, the square in column b and row 3.

One of the many green squares that we must count

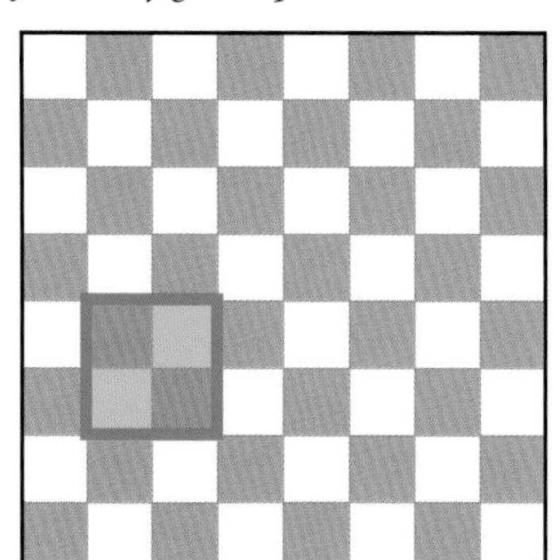

Inspecting the previous diagram reveals that any square other than those in column h or row 8 can serve as the lower-left corner of a 2×2 square. It follows there are forty-nine squares of size 2×2 in the diagram.

Imitating this logic for the case of 3×3 squares leads to the observation that any square except for those in rows 7 or 8, or those in columns g and h, can serve as the lower-left corner of a 3×3 square. Consequently, there are thirty-six such squares.

Perhaps the pattern is now clear? The number of squares of size $n \times n$, where n is some integer between 1 and 8 inclusive, is given by $(9 - n)^2$. For example, the number of 3×3 squares is $(9 - 3)^2 = 36$. It follows that we can count the total number of squares in the diagram by adding up the number of squares of size 1×1, size 2×2, size 3×3, and so on until we reach the lone 8×8 square that is the entire checkerboard:

$$8^2 + 7^2 + 6^2 + 5^2 + 4^2 + 3^2 + 2^2 + 1^2 = 204$$

The nice thing about this method is that it generalizes easily to checkerboards of arbitrary size. If we have a checkerboard of size $d \times d$, then the total number of squares it contains is given by (reversing the order from our previous example):

$$1^2 + 2^2 + 3^2 + \ldots + (d - 1)^2 + d^2.$$

As it happens, there is a well-known formula for the sum of the first d perfect squares. It is given by

$$1^2 + 2^2 + 3^2 + \ldots + (d-1)^2 + d^2 = \frac{d(d+1)(2d+1)}{6}.$$

For example, the sum of the first eight perfect squares is given by

$$\frac{8(9)(17)}{6} = 204.$$

Our second example involves a single-elimination tennis tournament. By single-elimination, we mean that as soon as a player loses a match he is eliminated from the tournament. The tournament continues until only one player remains. The question is: How many matches are played in the tournament?

Obviously that depends on how many people started in the tournament. Let us work our way in with some easy examples.

Puzzle 25: Tournament Matches.

Suppose sixty-four people compete in a single-elimination tennis tournament. What is the total number of matches to be played in the tournament? What if there were eighty people instead?

Let us consider a much larger tournament, say with 1,024 tennis players. Our first approach will be simple brute force, working through the series of matches that must be played. In the first round, 1,024 players will engage in 512 matches. In the next round, the remaining 512 players will engage in 256 matches. The third round will feature 128 matches, then 64 in the round after that, and so on. It is then a simple computation to show that the total number of matches is

$$512 + 256 + 128 + \ldots + 4 + 2 + 1 = 1{,}023.$$

Brute force, however, is not very illuminating. It also is not as helpful in cases where the number of players is not a power of 2. (Note that in our example we have $2^{10} = 1{,}024$). In such cases, certain rounds will feature an odd number of players, and we must assume the tournament organizers have worked out some system of byes to handle them. We can still ask how many matches will be played before a winner emerges. We assume only that we start with x players and keep playing until only one player remains.

Initially the situation appears hopeless. Since we know nothing about the number of players with which we began, there is no hope of using any sort of computational approach.

That, however, is not cause for despair. What if instead of counting the number of matches, we counted instead some other set that has a size equal to the number of matches? There is such a set, which you will realize for yourself as soon as you consider that every match produces one winner and one loser. Since there is only one winner at the end of the tournament, there must be $x - 1$ losers. And since every match produces only one loser, we find there must have been $x - 1$ matches played. This is precisely what we found in the special case where the number of players was 1,024, as well as in Puzzle 25 where that number was 64, since those numbers are both powers of 2.

The point? Just that counting problems sometimes require cleverness and ingenuity. We are going to need both in determining the number of Sudoku squares.

4.2 COUNTING SHIDOKU SQUARES

How can we use these principles to determine the number of Sudoku squares? Unfortunately, this is a much more difficult problem than the ones discussed so far.

Things start innocently enough: Clearly there are 9! ways of choosing an order for the numbers in the top row of any Sudoku square. After that things get complicated quickly, because of the block and column conditions. Given the first row, there are fewer than 9! ways of filling in the second row in a manner consistent with the rules of Sudoku. The third row is still more restricted, and subsequent rows are more restricted still. Good luck trying to keep track of everything.

Since counting Sudoku squares is a challenging problem, let us warm up with a simpler one. Instead of considering a 9×9 Latin square that is subdivided into 3×3 blocks, consider a 4×4 Latin square that is subdivided into 2×2 blocks, as in Puzzle 26. If we require the numbers 1–4 each to appear once in every row, column, and 2×2 block, then we will have a *Shidoku board*. A Shidoku puzzle is then a subset of a Shidoku board that can be completed in exactly one way.

Puzzle 26: Shidoku.

Fill in each grid so that every row, column, and block contains 1–4 exactly once. Compared to 9×9 Sudoku, these are pretty easy. If you have kids, here is their chance to write in your book!

4			
3	1		
		4	1
			2

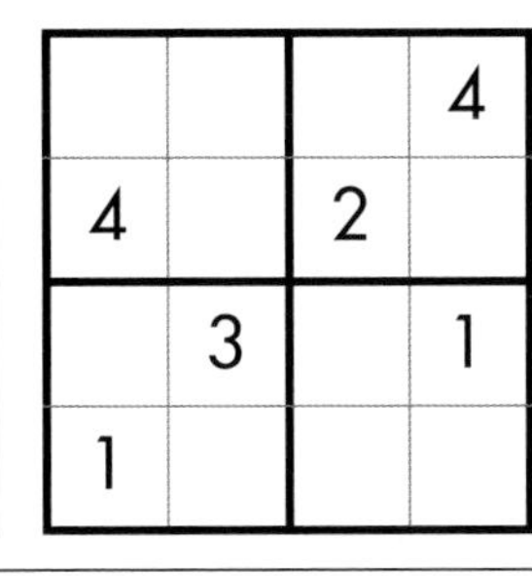

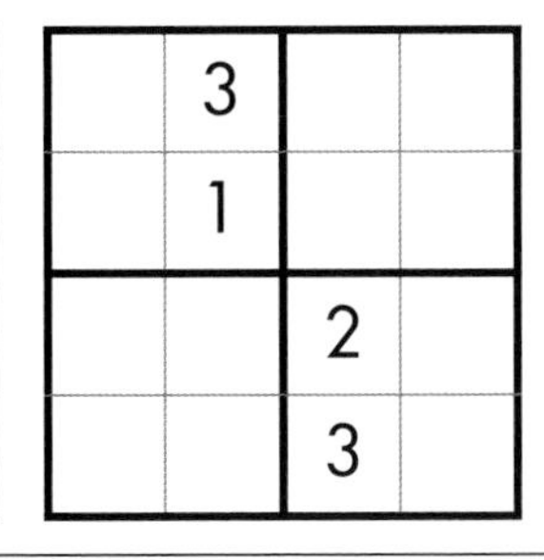

How many Shidoku squares are there? That is, starting from a blank grid, in how many ways can we place the numbers 1–4 into the grid so that each row, column, and 2×2 block satisfies the rules of Shidoku?

At a loss for how to proceed? Welcome to the world of research mathematics. That feeling of hopeless confusion, the uncertainty of how best to attack a problem, is one mathematicians experience through much of their professional lives. It can be frustrating, but it is motivating as well. The more opaque the problem, the greater the satisfaction when the light finally shines through.

We will begin by placing the entries of the first row, column, and 2×2 block of a Shidoku square in an "ordered" fashion.

Let us start with the upper-left 2×2 block of a Shidoku grid. We need to place the numbers 1–4 in this block, and our second counting principle tells us that there are $4! = 1 \times 2 \times 3 \times 4 = 24$ ways of doing so. One especially nice ordering is the following:

Shidoku grid with 'ordered' first block

1	2		
3	4		

Our problem is now reduced to counting the number of Shidoku squares whose upper-left 2×2 block has the pattern above. Though we will not pause to prove it here, it is clear that each of the twenty-four placements of numbers in the first block has the same number of completions. Let us call that number x. Filling in a Shidoku square can now be viewed as a two-step process: First we place 1–4 in the first block, and then we complete the rest of the square. By our first counting principle, the total number of Shidoku squares is then $24x$.

We now fill in the rest of the first row and the rest of the first column. We have two choices for completing the row, and two choices for completing the column. We will make the choice below in which the blue cells increase from left to right, and the green cells increase from top to bottom. (The sole alternative for the first row is to reverse the order of the digits 3 and 4 in the blue cells. Likewise, the sole alternative for the green cells would be to reverse the digits 2 and 4. Thus there are $2 \times 2 = 4$ ways that we could have chosen to order the last two rows and columns.)

Shidoku grid with 'ordered' first row, column, and block

1	2	3	4
3	4		
2			
4			

Now we need only count the number of Shidoku squares whose first row, column, and block are as above, and then multiply that number by $24 \times 2 \times 2 = 96$, to get the total number of Shidoku boards. We will refer to such squares as being *ordered.*

How do we complete our count? By playing Shidoku with our ordered squares, of course.

Puzzle 27: Ordered Shidoku Boards.

Find *all* of the possible ways of completing the ordered Shidoku grid above to make a valid Shidoku square. You should find that there are three such ways, so here are three grids for writing them out:

1	2	3	4
3	4		
2			
4			

1	2	3	4
3	4		
2			
4			

1	2	3	4
3	4		
2			
4			

Since there are exactly three ways of completing a Shidoku grid with ordered first row, column, and block, and since exactly ninety-six Shidoku squares are represented by each 'ordered' square, we have shown that the number of possible Shidoku squares is

$$96 \times 3 = 288.$$

Not a bad trick! Can we apply this method to the 9×9 case?

4.3 HOW MANY SUDOKU SQUARES ARE THERE?

Our task now is to extend the techniques of the previous section to the problem of counting Sudoku squares. This problem was first solved by Felgenhauer and Jarvis [23], and we will follow their proof.

To facilitate the discussion, we label the 3×3 blocks of a Sudoku square as follows:

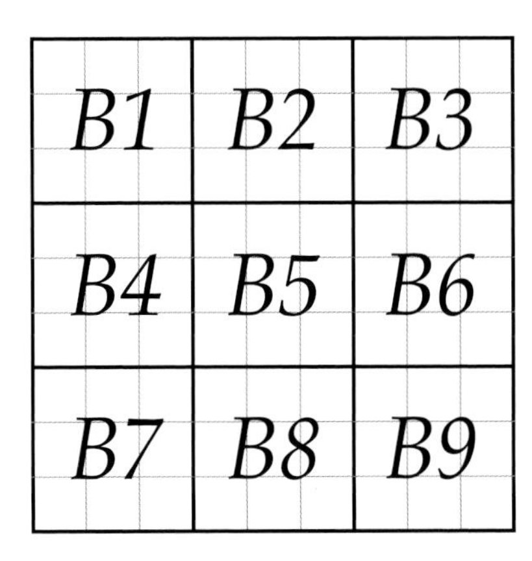

As with the Shidoku squares, we begin by relabeling the digits so that $B1$ has the form

1	2	3
4	5	6
7	8	9

($B1$ = the block above)

Any Sudoku square with $B1$ arranged in this way will be said to be in *standard form*. Notice that there are $9! = 362,880$ other Sudoku squares obtainable from a square in standard form by relabeling the cells. The number we seek is therefore the number of Sudoku squares in standard form multiplied by 9!.

We move now to $B2$ and $B3$. With $B1$ in standard form, in how many ways can we fill in $B2$ and $B3$? This is where things get tricky as compared to counting Shidoku squares. We have so many more possibilities to consider, you see.

Which numbers could go in the top row of $B2$? There are only three possibilities:

Case 1(a). The top row of $B2$ is some arrangement of the three numbers from row 2 of $B1$ (that is, 4, 5, 6 in some order).
Case 1(b). The top row of $B2$ is some arrangement of the three numbers from row 3 of $B1$ (that is, 7, 8, 9 in some order).
Case 2. The top row of $B2$ is some mixture of numbers taken from rows 2 and 3 of $B1$ (for example, 4, 5, 7 in some order or 5, 7, 9 in some order).

In both parts of case 1 there are only two possibilities for the digits appearing in the top rows of blocks $B2$ and $B3$. First, suppose the top row of block $B2$ is a permutation of $\{4, 5, 6\}$; then the top row of block $B3$ must be some permutation of $\{7, 8, 9\}$. In fact, we can say more; in this situation, we also know how to fill in the second and third rows of $B2$ and $B3$, up to permutation.

The top three rows of the Sudoku square will look like this:

1	2	3	{4,5,6}	{7,8,9}
4	5	6	{7,8,9}	{1,2,3}
7	8	9	{1,2,3}	{4,5,6}

In each of the six rows of $B2$ and $B3$ we can choose any of the 3! permutations of the three entries. In other words, the first row of $B2$ could be 4, 5, 6 or 4, 6, 5 or 6, 5, 4 or three others besides. Things are similar when we are in the second scenario of case 1, in which case, the top row of block $B2$ is a permutation of $\{7, 8, 9\}$. Then the top row of $B3$ must correspondingly be a permutation of the remaining numbers $\{4, 5, 6\}$.

This means that in both types of case 1 situations, filling in the rows of blocks $B2$ and $B3$ is a six-step process in which each of the six mini-rows can be completed in

3! different ways, for a total of $(3!)^6$ ways of filling in the rows in each of the two case 1 situations. Our first counting principle now dictates that there are

$$2 \times (3!)^6 = 93{,}312$$

ways of filling in blocks $B2$ and $B3$ in case 1.

Case 2 squares are more difficult, but a concrete example ought to help. In case 2 the entries in the top rows of $B2$ and $B3$ are a mixture of the numbers from rows 2 and 3 of $B1$. One possibility is that the top row of $B2$ is a permutation of $\{4, 5, 7\}$ and the top row of $B3$ is a permutation of $\{6, 8, 9\}$. Counting the number of possible completions is a little trickier in this case. We will let you have a go at it first:

Puzzle 28: The Top Three Rows.

The 3×9 grid below represents the top three rows of a Sudoku square. Suppose the first block is filled in as shown, the first row of $B2$ is some permutation of $\{4, 5, 7\}$, and the first row of $B3$ is some permutation of $\{6, 8, 9\}$. Under these conditions, in how many possible ways can the first three rows be completed? The solution is in the text below.

1	2	3	{4,5,7}			{6,8,9}		
4	5	6						
7	8	9						

Let us work through the solution. There are 3! ways of ordering the $\{4, 5, 7\}$ in the first row of $B2$, and 3! ways of ordering the $\{6, 8, 9\}$ in the first row of $B3$. Considering the fact that each digit can appear only once in each row, column, and block, the first three rows must look something like this (again using the curly-bracket notation to denote that the numbers appear in some unspecified order):

1	2	3	{4,5,7}	{6,8,9}
4	5	6	{8,9,?}	{7,?,?}
7	8	9	{6,?,?}	{4,5,?}

We need to put the numbers 1, 2, and 3 into the cells marked with question marks. Whatever goes into the blue question mark in the second row must also go into the blue question mark in the third row, while the remaining two digits must go into the cells with red question marks. There are three ways of choosing which of the digits 1, 2, and 3 is set aside for the blue question mark.

Therefore, given the $\{4, 5, 7\}$ and $\{6, 8, 9\}$ in the first row, filling in the first three rows requires two further steps:

1. Place one of the numbers 1, 2, 3 in the cells indicated by the blue question mark. Having made that choice, the numerical values of the red question

marks are now determined. There are three ways of carrying out this step – one way for each choice of blue question mark.

2. We must now choose an ordering for each of the six rows in $B2$ and $B3$. In each row we have 3! orderings from which to choose. Thus, ordering all the rows involves making six choices, each of which can be made in 3! many ways. It follows that this step can be carried out in $(3!)^6$ many ways.

We conclude that there are $3 \times (3!)^6$ ways of filling in the first three rows in this scenario.

Do all case 2 scenarios follow this pattern? Actually, yes. In any case 2 scenario, we must fill the first row of $B2$ with three digits from 4–9, without choosing either $\{4, 5, 6\}$ or $\{7, 8, 9\}$. Listing all of the ways of doing this is not difficult: For the first row of $B2$ we could choose $\{4, 5, 7\}$ as above, or $\{4, 5, 8\}$, $\{4, 5, 9\}$, or $\{5, 6, 7\}$, and so on. If you list all the possibilities you will find that there are eighteen ways of choosing three digits for a case 2 first row of $B2$.

Each of these choices involves either two digits from the second row of $B1$ and one digit from the third row of $B1$, or vice versa. This means that the general pattern we investigated in Puzzle 28 will hold for every case 2 possibility. Therefore there are a total of

$$18 \times 3 \times (3!)^6 = 2{,}519{,}424$$

ways of filling in blocks $B2$ and $B3$ in case 2.

Combining this with our findings for case 1 shows that there are

$$93{,}312 + 2{,}519{,}424 = 2{,}612{,}736$$

ways of filling in the first three rows of a Sudoku square whose first block $B1$ is in standard form.

Considering that there are 9! ways of filling in the first block, and that each of them is equivalent to the case we examined above, we see that there are

$$2{,}612{,}736 \times 9! = 948{,}109{,}639{,}680$$

ways of filling in the first three rows of a Sudoku board. Alas, even after all this work there are still six rows to go!

4.4 ESTIMATING THE NUMBER OF SUDOKU SQUARES

To this point, we have used standard counting techniques, and it would be nice to be able to finish the problem in the same way. How far can we get with this approach? The following heuristic argument is presented by Felgenhauer and Jarvis [23], where it is attributed to Kevin Kilfoil. By a 'heuristic' argument, we mean one that comes very close to the right answer while not quite being entirely correct.

The heuristic argument goes like this: We know how many ways we can fill in the entries of the first three rows of a Sudoku square. The same argument applies to rows 4–6 and 7–9. Consequently, the number of ways of filling in a 9×9 grid so that each block and each row contains the digits 1–9 exactly once is given by $(948{,}109{,}639{,}680)^3$. Notice that this number does not take into account any condition on the columns.

The number of ways a 9×9 grid could satisfy both the block condition and the *column* condition is also given by $(948{,}109{,}639{,}680)^3$. We simply use the same argument with our heads turned ninety degrees.

Now think for a moment about how many ways a 9×9 square could satisfy just the block condition. Our only concern here is that the numbers 1–9 appear exactly once in each block. There are 9! ways of placing the numbers 1–9 in a block, and thus $(9!)^9$ different ways of filling in a 9×9 square with each number appearing exactly once in each block.

Putting together these facts, we see that the probability that a 9×9 block-compliant square is also row-compliant is given by

$$k = \frac{(948{,}109{,}639{,}680)^3}{(9!)^9}.$$

Of course the number k is also the probability that a 9×9 block-compliant square is also column-compliant.

There is a basic principle of probability that if two events are independent of one another, then the probability that both occur is found by multiplying the probabilities of the individual events. Now *if* the row and column conditions were independent, we could use this principle to compute the number of Sudoku squares. That is a big "if," but let us roll with it for a moment.

Imagine selecting a random 9×9 square that satisfies the block condition. The probability that it is row-compliant is k. The probability that it is column-compliant is also k. Therefore, the probability that it is both row-compliant and column-compliant is given by k^2, so long as these probabilities really are independent.

Continuing to assume independence of the row and column conditions, we could compute the total number of Sudoku squares by multiplying the total number of block-compliant squares by the probability that a randomly selected such square is both row-compliant and column-compliant. That is, the total number of Sudoku squares would be given by

$$(9!)^9 \times k^2 = (9!)^9 \times \left(\frac{(948{,}109{,}639{,}680)^3}{(9!)^9} \right)^2 \approx 6.6571 \times 10^{21}.$$

As we shall soon see, this number is strikingly close to the correct answer. That it cannot actually be the correct answer becomes clear once you notice it is not an integer. Its decimal form is approximately

$$6{,}657{,}084{,}616{,}885{,}512{,}582{,}463.49.$$

The flaw in the argument is the assumption that the row and column probabilities are independent. Sadly, they are not. Knowing that a randomly selected 9×9 square satisfies the row condition affects the probability that it also satisfies the column condition. Clever as this argument is, it is not quite correct.

However, the work above does provide a good estimate of the exact answer. Since the row and column conditions are dependent on one another, we expect the actual number of Sudoku squares to be somewhat larger than what we just computed. As we will see, that is precisely what Felgenhauer and Jarvis [23] discovered.

4.5 FROM TWO MILLION TO FORTY-FOUR

To find the exact answer we need a different approach. There is one natural possibility that comes to mind. We could try taking each of those 2,612,736 configurations for the top three rows with ordered first block, and simply play Sudoku with them to find their possible completions. This is essentially what we did in the Shidoku case.

Unfortunately, in the 9×9 case, doing such a thing by hand is too cumbersome. A computer, on the other hand, would fare better. We could have the computer try every possible way of filling in the rest of the square and keep track of the ones that satisfy the rules of Sudoku.

The trouble is that the numbers we are talking about are too large even for a computer. If we are going to complete our count, we need to reduce the number of cases to be checked. Happily, there are several ways of doing that.

As usual, a specific example will help illustrate our strategy. Below is one possible arrangement for the first three rows of a Sudoku square in standard form. Let us call this configuration 1. (We highlight two columns on which we will focus shortly.)

Configuration 1

1	2	3	5	7	6	4	8	9
4	5	6	8	1	9	3	2	7
7	8	9	2	3	4	1	6	5

Now consider the following arrangement, obtained by swapping the fifth and sixth columns of configuration 1. We will call this new arrangement configuration 2.

Swapping columns 5 and 6 gives us configuration 2

1	2	3	5	6	7	4	8	9
4	5	6	8	9	1	3	2	7
7	8	9	2	4	3	1	6	5

The number of ways of completing configuration 1 to a Sudoku square is the same as the number of ways of completing configuration 2.

Here is why: If S is a completion of configuration 1 to a Sudoku square, then reversing the fifth and sixth columns of S will be a completion of configuration 2. Likewise, any completion of configuration 2 also leads to a completion of configuration 1 by reversing the columns. In this way, we have a perfect pairing between completions of configuration 1 with completions of configuration 2. It follows that the two configurations have the same number of completions.

We could also permute entire blocks. Starting from configuration 2:

Configuration 2, this time with blocks shaded

1	2	3	5	6	7	4	8	9
4	5	6	8	9	1	3	2	7
7	8	9	2	4	3	1	6	5

we can obtain an essentially equivalent configuration by reversing blocks $B2$ and $B3$:

Swapping second and third blocks gives us configuration 3

1	2	3	4	8	9	5	6	7
4	5	6	3	2	7	8	9	1
7	8	9	1	6	5	2	4	3

This new configuration can be completed to a Sudoku square in precisely the same number of ways as configurations 1 or 2.

In general, in any configuration of the first three rows, we can swap columns within a block, or swap the blocks themselves, and end up with a new configuration with the same number of completions. We do not want to waste time counting the number of completions separately for two essentially equivalent boards. Luckily, with column and block swaps, each configuration of the first three rows can be turned into a configuration whose first row is what we shall call "ordered," as follows:

1. If necessary, permute the columns within $B2$ and $B3$ so that the entries in the first row of each block are in increasing order.
2. If necessary, now exchange $B2$ and $B3$ so that the upper left entry of $B2$ is smaller than the upper left entry of $B3$.

Our example above happened to illustrate this exact process; we swapped columns and blocks to turn our original configuration 1 into an essentially equivalent configuration whose first row $(1, 2, 3)$, $(4, 8, 9)$, $(5, 6, 7)$ was 'ordered' in the sense discussed above. This configuration represents some of the possible boards in our case 2 scenario from before.

If we follow the two-step ordering process above for *any* case 1 situation, we end up with just *one* possible way we could have ordered the first row of nine numbers:

Ordered first row in a case 1 situation

1	2	3	4	5	6	7	8	9
4	5	6	{7,8,9}			{1,2,3}		
7	8	9	{1,2,3}			{4,5,6}		

Note that we have reduced case 1 from $2 \times (3!)^6 = 93{,}312$ possible configurations to just $(3!)^4 = 1{,}296$ configurations; this is an improvement by a factor of $2 \times (3!)^2 = 72$.

Improvement by a factor of 72 also happens in general: Since there are six ways of permuting the columns in $B2$ and six ways of permuting the columns in $B3$, step 1 implies there are 36 configurations having as many completions as the one above. Step 2 doubles that number to 72. Instead of having to check more than 2.6 million configurations for the first three rows, we now only have to check

$$\frac{2{,}612{,}736}{72} = 36{,}288.$$

A substantial improvement, but still too many to be practical.

It turns out, however, that there are many additional reductions to consider, some of them too complex to present here. The strategy, however, remains the same. We break up our large set of configurations into smaller sets in such a way that within each set, each configuration has the same number of completions to a full Sudoku square. It then suffices to check just one representative from each of the small sets.

By using this strategy, Felgenhauer and Jarvis [23] managed to show that instead of checking over 2.6 million possibilities, it sufficed to check a cleverly chosen 44. That number is computationally feasible!

4.6 ENTER THE COMPUTER

It is still far too many to check by hand, however. To count the number of Sudoku squares, we need to find all possible completions of each of the forty-four representative configurations of the first three rows. Try to complete the remaining six rows for any of the examples shown above, and you will see that there are going to be a great many completions indeed. Also, it is important to note that these forty-four configurations could have different numbers of possible completions to full Sudoku squares.

In the corresponding 4×4 Shidoku case, the completion process is far simpler, although the following puzzle should give you an idea of what is required in the 9×9 case.

Puzzle 29: Completing Shidoku.

Consider the following two partially completed Shidoku boards. Find the number of possible completions to full Shidoku boards in each case. (Notice that we have ordered the first block and first row, and then completed the second row, in a way analogous to what is described for Sudoku.)

1	2	3	4
3	4	1	2

1	2	3	4
3	4	2	1

Felgenhauer and Jarvis needed a computer to grind out the final answer in the 9×9 case. They found all the possible completions of each of their forty-four configurations to show that there are

$$6{,}670{,}903{,}752{,}021{,}072{,}936{,}960$$

possible Sudoku squares. Problem solved!

This number is approximately 6.67 sextillion, which is more than the number of stars thought to be in the known universe, or the number of grains of sand on all the Earth's beaches. It would seem the newspapers will not be running out of puzzles any time soon.

Does it feel like cheating that we used a computer to do the heavy lifting? Mathematics is supposed to be an exercise in pure logic in which things get proved to an absolute certainty. Trusting a machine to tell us what is and is not true is hardly in the spirit of things.

Most mathematicians would agree. The issue is not that we are worried that the computer is defective in some way. That the computer says a statement is true is enough to convince almost everyone that it is, indeed, true. Rather, the problem is that the computer provides little insight into *why* a result is true.

Remember in the last chapter when we said a good proof does more than simply establish the truth of a theorem? It is supposed to clarify, not just verify. The computer verifies without clarifying.

But what is the alternative? If we had a purely logical argument for counting the number of Sudoku squares, we would have shown it to you. To date, no one has devised such a thing; the first person who does will be a star among those who enjoy Sudoku. For the moment, our choices are accepting the computer proof or wallowing in ignorance. Verification may not be everything, but it is hardly nothing.

Computers have been playing an ever-increasing role in mathematics ever since Appel and Haken proved the Four-Color Theorem in 1976 [4]. Imagine a map of the United States, or perhaps of the countries in Europe. It is desirable that each state or country be colored in such a way that neighboring regions have different colors. The question is how many colors are needed for this purpose. It turns out that four colors suffice for any such map. This problem has a history going back to the mid-nineteenth century.

Appel and Haken [4] followed a strategy nearly identical to the one we used here. They reduced the problem to showing that every map was effectively equivalent to one among a small number of test cases. They then used the computer to verify that each of their test cases could be colored with four colors. This was the first major theorem to be proved with the assistance of a computer. Since then, as computing power has increased and has become more readily available, the computer has become an indispensable tool in many branches of mathematical inquiry. (We shall discuss this further in Chapter 7).

It is not our purpose here to hash out the philosophical issues raised by computer-assisted proofs. We have a different point in mind. It is that mathematics is not a static discipline. It is not something going on solely in ivory towers, entirely divorced from the surrounding world. The topics mathematicians choose to study and the techniques used for solving problems change with the times just as surely as does art and literature and science.

4.7 A NOTE ON PROBLEM-SOLVING

Which brings us to our final point. We worked pretty hard to count those Sudoku squares, and we did not even present all of the details. At first it was unclear how to get started; having started, it was often unclear how to proceed. At one point, we followed a promising lead only to have it peter out to a dead end.

That is how the game is played. Sometimes a problem yields to a flash of insight and a cry of "Eureka!" but usually the solution only comes after much wandering in the wilderness. Success in mathematics has far more to do with persistence and hard work than it does with raw genius. Most people, upon failing to see a solution within a few seconds, move on to something else and dismiss the problem as unimportant. A mathematician reaches the same point and sees the battle as well and truly joined. He is not about to have the problem defeat him in one-on-one combat. The harder the problem, the more satisfying is its eventual fall.

Furthermore, the journey is usually more important than the answer. This point is often ignored by students who have been raised with the idea that "getting the right answer" is all that matters. When you think about it, though, the revelation of the precise number of Sudoku squares was by far the least interesting part of this chapter. Turns out the number is on the order of 10^{21}. So what? If it had been on the order 10^{22} or 10^{18} would anyone have cared? The interesting part was the reasoning that went into the solution. From our proof we not only learn the precise number of Sudoku squares, an amusing but largely irrelevant bit of trivia, but we also learn something about how to solve difficult counting problems. *That* is something of real significance.

As a final example, consider again the game of Sudoku. The fun is not in discovering which particular Sudoku square happens to be the solution to the puzzle you are playing. Instead, the fun is in the journey. And after all the work we did in this chapter, we deserve a little extra fun:

Puzzles 30 and 31: Sudoku Reward.

Fill in each grid so that each row, column, and block contains each of the numbers 1–9 exactly once. The first puzzle is a breezy trip to the beach, but the second one should take you on more of a journey.

		6						
3							9	7
			4			2	8	3
		5		8	9			
9								1
			2	1		4		
6	1	3			5			
8	5							2
						5		

2	1					6		
8			9					
			8					4
				3	7		9	
	8		5	4				
3					6			
					1			9
		4					1	5

5

Equivalence Classes

The Importance of Being Essentially Identical

Perhaps there is something bothering you about the last chapter. Have a look at these two Sudoku squares:

6	4	5	7	8	9	1	2	3
7	8	3	2	1	6	4	5	9
2	1	9	4	5	3	6	7	8
9	6	1	8	7	2	5	3	4
5	3	7	9	6	4	8	1	2
8	2	4	1	3	5	9	6	7
1	7	2	6	4	8	3	9	5
3	9	8	5	2	1	7	4	6
4	5	6	3	9	7	2	8	1

7	5	6	8	9	1	2	3	4
8	9	4	3	2	7	5	6	1
3	2	1	5	6	4	7	8	9
1	7	2	9	8	3	6	4	5
6	4	8	1	7	5	9	2	3
9	3	5	2	4	6	1	7	8
2	8	3	7	5	9	4	1	6
4	1	9	6	3	2	8	5	7
5	6	7	4	1	8	3	9	2

To a casual glance they appear to be different. Look more closely, however, and you will notice that each cell in the second square is one more than the corresponding cell in the first square. That is, every occurrence of 1 in the first square has been replaced by 2 in the second square. Every occurrence of 2 has been replaced by 3, and so on with every occurrence of 9 being replaced by 1.

Our count in the previous chapter treated these as different squares. Somehow that does not seem quite right. Since we have previously emphasized that the specific symbols used to fill in the cells are irrelevant, we have not really changed the puzzle just by renaming the symbols. This raises a question: How many *fundamentally different* Sudoku squares are there?

Which, in turn, raises the question of precisely what we mean by "fundamentally different." Both questions can be answered, but we first need to lay some groundwork.

5.1 THEY MIGHT AS WELL BE THE SAME

Puzzle 32: Days of the Week.

If today is a Tuesday, what day of the week will it be 7,000 days from now? How about 9,326 days from now? You should be able to answer both questions in your head! The solution is in the text below.

It is customary for the students in a given high school to be partitioned into freshmen, sophomores, juniors, and seniors. This can be very useful in many situations. For example, there might be a test that all seniors have to pass in order to graduate. If we then want to know whether a given student is required to take the exam, we need only inquire as to the year he is in. We do not care about the student's name, home address, hair color, or any of the manifold other things that differentiates him from his fellow students. As far as taking the test is concerned, one senior is the same as any other.

Biologists divide the animals they find in nature into separate species. Two animals are placed in the same species if they belong to an interbreeding community of other animals. Human beings are one species, while dogs are another. In making these divisions the biologists are not saying that all people or all dogs are the same. Instead they are simply making divisions among animals that are useful in a great many circumstances. For certain purposes it is convenient to treat all dogs as essentially equivalent.

If x and y are two whole numbers, will their sum be even or odd? To answer that question it is not necessary to know the precise values of x and y. You only need to know whether x and y are themselves even or odd. You know that the sum of two even numbers is even, the sum of two odd numbers is also even, but the sum of an even and an odd number is odd. Therefore, if I tell you that x and y are both even then you know enough to conclude the sum is even. The precise even numbers represented by x and y are entirely irrelevant.

These are all examples of *equivalence*. Informally, mathematicians describe two objects as equivalent if they might as well be the same. That is, they possess some property in common that allows us to treat them as identical for certain purposes. The basic principle presents itself every time you divide a large and diverse collection of objects into smaller, more manageable sets. If we are to answer the questions asked in the introduction to this chapter, we will need a clearly defined notion of equivalence for Sudoku squares. We have already seen one small part of such a definition: Two Sudoku squares are equivalent if one can be obtained from the other by systematically renaming the symbols. As we shall see, however, our definition will need to contain many other items as well.

What has this to do with the puzzle at the start of the section? We divide the days of the week into Mondays, Tuesdays, Wednesdays, and so on. We know that if today is Monday, then every 7 days from now it will be Monday again. Thus, after 7 days it will be Monday, and likewise for 14 days, 21 days and so on. It follows that if I want to know the day of the week some number of days hence, all that matters is the remainder that number leaves when divided by 7. The number 7,000 is obviously a multiple of seven. It follows that 7,000 days from now will be another Tuesday. If I divide 9,326 by 7 I obtain a remainder of 2. It follows that 9,326 days from now will be a Thursday.

5.2 TRANSFORMATIONS PRESERVING SUDOKUNESS

When are two Sudoku squares equivalent?

We will approach this question by asking something else. Given a completed Sudoku square, what sorts of transformations can we aaply without disrupting its Sudokuness? We have already seen one example. If we systematically relabel the cells the result is another Sudoku square. We now seek a list of other such transformations. You might enjoy trying to come up with some yourself before reading on.

What would happen if we simply switched two cells possessing different digits? Would the result be another Sudoku square? Clearly not. Suppose x and y are different digits appearing in different rows (the case of different columns is equivalent). After the switch, we will have two occurrences of x in one row and two occurrences of y in a different row. No good.

How about switching two rows (or columns)? That is a tougher question, for a reason illustrated by the following diagrams. For convenience, we shall use Shidoku squares here, though the same principles hold true for Sudoku.

Shidoku

1	2	3	4
3	4	1	2
2	3	4	1
4	1	2	3

still Shidoku

3	4	1	2
1	2	3	4
2	3	4	1
4	1	2	3

not Shidoku

2	3	4	1
3	4	1	2
1	2	3	4
4	1	2	3

The first is a Shidoku square in ordered form as described in the previous chapter. The second was obtained from the first by switching the first two rows. It is still a Shidoku square. The third square was obtained by switching row 1 with row 3. Alas, it is no longer a Shidoku square. It would seem that sometimes reversing two rows produces another valid square, and sometimes it does not.

Perhaps you have already noticed the problem. We will need some new terminology to express it properly. Let us refer to a set of three horizontally adjacent blocks as a *band*, and a set of three vertically adjacent blocks as a *pillar*. Using the notation from the previous chapter, we could say that blocks $\{B1, B2, B3\}$ taken together form a band, as do $\{B4, B5, B6\}$ and $\{B7, B8, B9\}$. The pillars are then given by blocks $\{B1, B4, B7\}$, $\{B2, B5, B8\}$, $\{B3, B6, B9\}$.

Bands

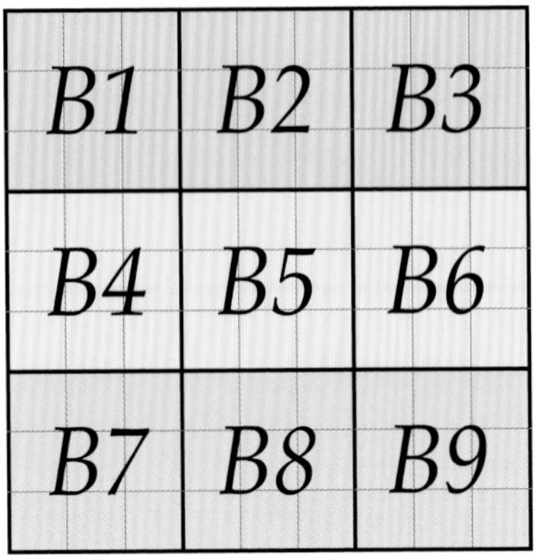

B1	B2	B3
B4	B5	B6
B7	B8	B9

Pillars

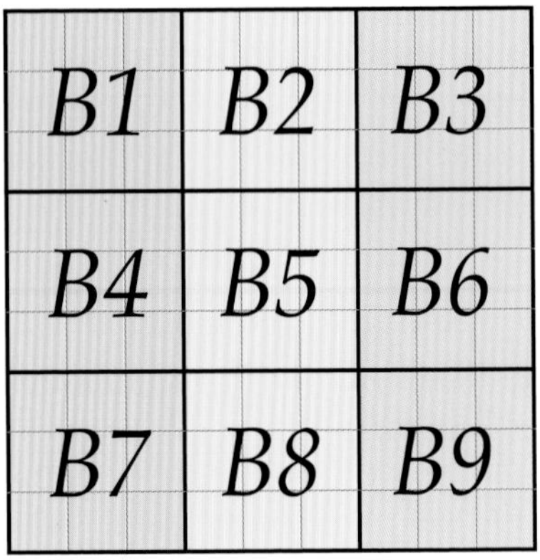

B1	B2	B3
B4	B5	B6
B7	B8	B9

Using this language, we can say that switching two rows within the same band (or two columns within the same pillar) will be a valid transformation. Switching rows and columns from different bands or pillars is not acceptable. More generally, we can permute the rows within a band or the columns within a pillar without losing our Sudokuness. Since these statements are fairly simple to prove, we will leave you to think about them for yourself.

A variation on this theme involves permuting the bands, or pillars, themselves. These transformations also produce valid squares and therefore can be added to our list.

Have we now listed all of our transformations? Not quite! In the following diagram, the first Shidoku square is the same one we used before. Should the other two squares be considered equivalent to the first?

Shidoku

1	2	3	4
3	4	1	2
2	3	4	1
4	1	2	3

still Shidoku

4	2	3	1
1	3	4	2
2	4	1	3
3	1	2	4

still Shidoku

3	1	2	4
2	4	1	3
1	3	4	2
4	2	3	1

They certainly were not obtained just by permuting rows, columns, bands, or pillars. We did, however, obtain them through simple processes. The second square was obtained from the first by rotating clockwise through 90 degrees. The third was obtained by reflecting along the diagonal leading from the upper-right corner to the lower-left corner. In fact, any such rotation or reflection through an axis of symmetry will lead to another valid square. We can rotate through 90, 180, or 270 degrees (or 0 degrees, for that matter). As for reflections, notice that there are four axes of symmetry from which to choose. The two main diagonals do the trick, as does the horizontal line through the middle of row 5, and the vertical line through the middle of column 5, as shown below.

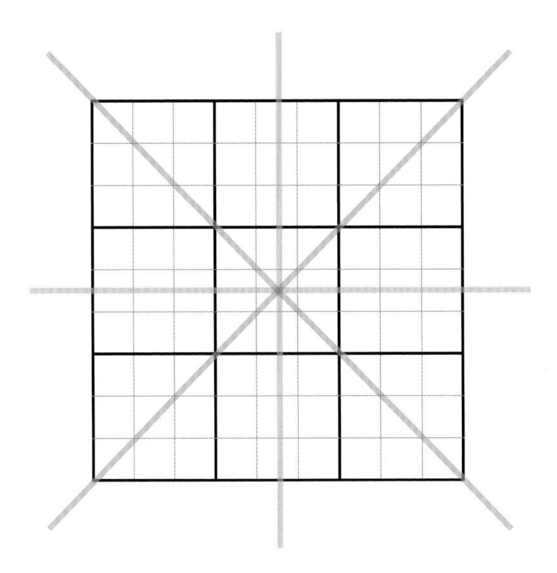

That makes four rotations and four reflections to add to our list.

Gathering everything together leads to the following list of valid transformations:

1. Relabeling the digits.
2. Permuting the rows in a band or the columns in a pillar.
3. Permuting the blocks in a given band or pillar.
4. Any rotation or reflection.

A fine list. Is it complete? It is indeed, though that fact is not so easy to prove. If you would like to see the details, you should consult the paper by Adler and Adler [3]. Any transformation of our square that preserves Sudokuness can be expressed as a combination of the items on the list.

We can now give a precise definition of what it means to say that two Sudoku squares are fundamentally different. It means that there is no combination of items on the list allowing us to transform one into the other.

5.3 EQUIVALENT SHIDOKU SQUARES

Let us pause to consider how these ideas relate to the Shidoku squares we counted in Chapter 4. We found there were 288 Shidoku squares. To establish this number, we first defined a reasonable notion of an "ordered Shidoku square." We then found that there were three ways of completing these ordered squares, as shown below:

1	2	3	4
3	4	1	2
2	1	4	3
4	3	2	1

1	2	3	4
3	4	1	2
2	3	4	1
4	1	2	3

1	2	3	4
3	4	2	1
2	1	4	3
4	3	1	2

The final step was to argue that each of these three squares was equivalent to 96 others (including itself), for a total of 288 squares. At this point, we can be certain that there are at most three fundamentally different Shidoku squares.

It turns out, however, that we did not take advantage of all possible equivalences. Have a look at the third square. It turns out that we can apply two of the transformations on our list to transform it into the second square. First, we reflect along the diagonal going from the upper left to the lower right. This has the effect of turning the first row into the first column, the second row into the second column, and so on. (If you are familiar with matrices, then you might recognize this operation as taking the transpose.) After carrying out this reflection, we then switch the 2s and the 3s. The result is shown below:

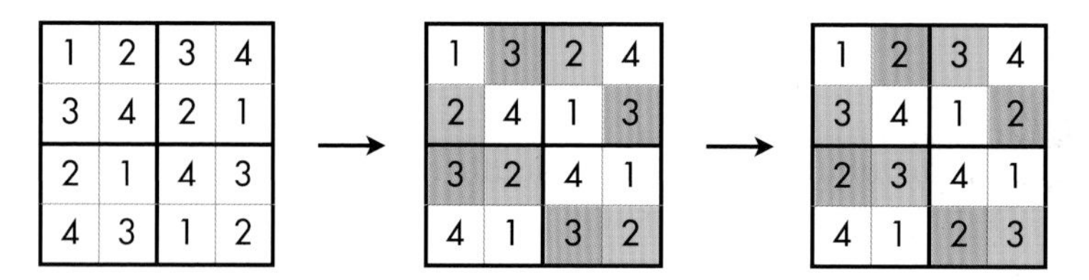

It would seem that while there are 288 different Shidoku squares, there are a mere 2 that can be described as fundamentally different.

Puzzle 33: Shidoku Equivalence.

We have just seen that of the 288 possible Shidoku squares, 96 of them are equivalent to the square on the left below, and 192 of them are equivalent to the square on the right.

Type 1 Shidoku

1	2	3	4
3	4	1	2
2	1	4	3
4	3	2	1

Type 2 Shidoku

1	2	3	4
3	4	1	2
2	3	4	1
4	1	2	3

Use Shidoku symmetries to determine which of the following boards are type 1 boards and which are type 2.

3	1	2	4
4	2	1	3
1	3	4	2
2	4	3	1

1	4	2	3
2	3	4	1
3	2	1	4
4	1	3	2

2	3	4	1
1	4	2	3
3	2	1	4
4	1	3	2

5.4 WHY THE NATURAL APPROACH FAILS

Now what?

We could try a direct approach to counting our fundamentally different squares. Choose a random Sudoku square. Call it S. Suppose we knew that S was equivalent to precisely x other squares. In that case, could we not simply take the total number of Sudoku squares and divide by x to obtain the number of fundamentally different squares? Sadly, no. That approach works only if x is independent of the particular square we chose. As we saw even in the 4×4 case, the number x of Sudoku squares that are equivalent to any given Sudoku square depends greatly on what Sudoku square we consider. Let's look more closely at what happens in the 9×9 case.

You are no doubt familiar with the idea of *symmetry*. Informally, an object is symmetric if it looks the same on one side as it does on the other. The human face is symmetric with respect to a vertical line drawn vertically through its center, for example. That vertical line is referred to as an *axis of symmetry*.

More generally, a symmetry of an object is any transformation that leaves some property of interest essentially unchanged. For the human face, the transformation involves reflecting across the axis of symmetry. Imagine a standard smiley face drawn on a piece of paper. If you place a thin, one-sided mirror along the axis, then the reflected half-face looks identical to the part of the face on the other side of the mirror.

Return now to our list of Sudoku-preserving transformations and notice that there is a crucial difference between type 1 transformations and types 2–4. Type 1 transformations are the only ones that involve changing the values found within the individual cells. The other transformations are actually just permutations of the cells (leaving unchanged the digits within the cells). For that reason, we will separate out the type 1 transformations. We'll begin by saying that two squares are equivalent, which is to say identical for our purposes, if they differ only by a type 1 transformation. With that definition in mind, we can refer to the other transformations as symmetries on the individual cells. In other words, they are permutations of the cells that preserve Sudokuness.

For convenience, from now on we will refer to a type 1 transformation as a *relabeling*, and other types of transformations as *permutations*.

Now we can explain the problem with the natural approach. There is no guarantee that applying a given symmetry to a square will produce a new square. Have a look at these two squares from Russel and Jarvis [36]:

1	2	4	5	6	7	8	9	3
3	7	8	2	9	4	5	1	6
6	5	9	8	3	1	7	4	2
9	8	7	1	2	3	4	6	5
2	3	1	4	5	6	9	7	8
5	4	6	7	8	9	3	2	1
8	6	3	9	7	2	1	5	4
4	9	5	6	1	8	2	3	7
7	1	2	3	4	5	6	8	9

7	4	8	5	2	9	6	3	1
1	9	6	4	3	8	5	7	2
2	5	3	6	1	7	9	8	4
3	6	9	7	4	1	8	2	5
4	1	7	8	5	2	3	9	6
5	8	2	9	6	3	1	4	7
6	2	1	3	9	4	7	5	8
8	3	5	2	7	6	4	1	9
9	7	4	1	8	5	2	6	3

The second square is obtained from the first by rotating ninety degrees clockwise. We will now carry out a relabeling on the second square. Permute the digits according to the following rule:

$$1 \to 3 \to 9 \to 7 \to 1.$$

This means that every occurrence of 1 should be changed to 3, every occurrence of 3 should be changed to 9, and so on with every occurrence of 7 being replaced by 1. For the remaining digits follow the rule

$$2 \to 6 \to 8 \to 4 \to 2.$$

Leave the 5s alone! Carry out this permutation and a remarkable thing happens. You are right back to the original square.

Do you see the problem? You do not get as many essentially identical squares as you think. In our example, rotation through ninety degrees failed to produce a new square, since the rotated square was equivalent to what we started with, after a relabeling. For other squares that does not happen. We will have to work harder to complete our count.

5.5 GROUPS

Do not despair! We are not yet out of tricks.

We have noted that all transformations of types 2–4 are permutations of the eighty-one cells of a square. If we carry out several of these permutations in succession, the result is yet another valid permutation. When we perform two or more transformations in succession in this manner, we will say we have *composed* them. We can think of composition as a binary operation on the set of symmetries, just like addition and multiplication are binary operations on the set of integers. Let us consider how these ideas play out in a simpler setting.

Imagine that you have an equilateral triangle. With a little trial and error, you will find that there are six symmetries. You can rotate through 120 degrees or 240 degrees. You could also do nothing, which shall be referred to as the *identity* transformation. There are three axes of symmetry to reflect across, one axis through each vertex bisecting the opposite side.

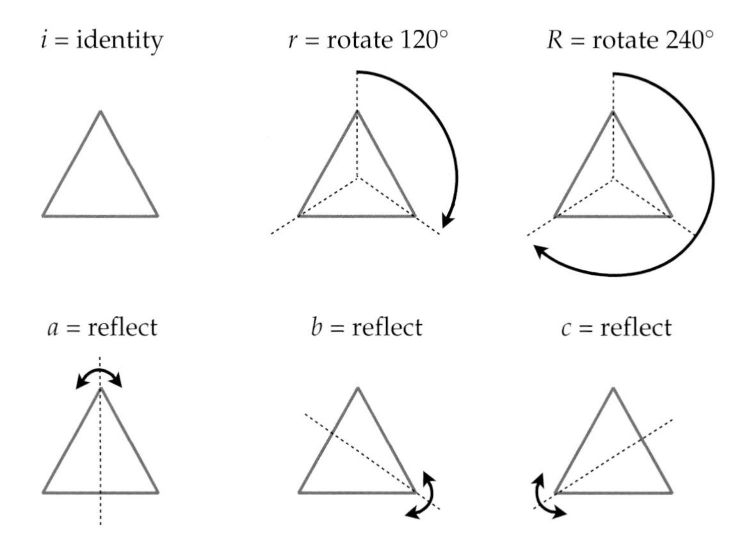

Let us see what happens when we compose various symmetries. In the following diagram, we went from the first triangle to the second by rotating 120 degrees clockwise, which we denote by r. To get to the third triangle, we then did the reflection a over the axis passing through the upper-most vertex. The sides of the triangle are colored so that you can follow the transformations.

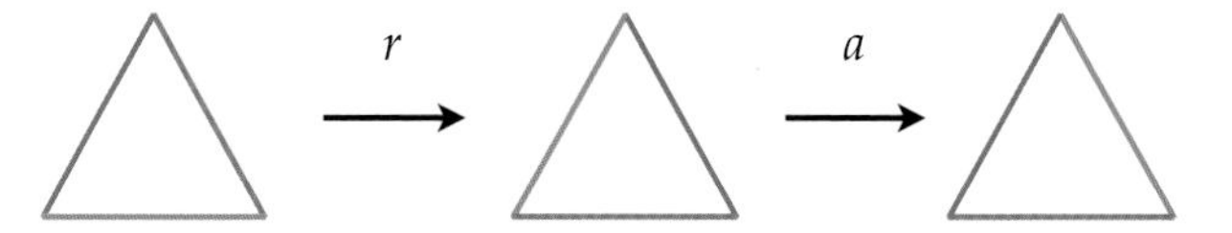

If you look carefully, you will notice that the third triangle could have been obtained from the first by just one symmetry, namely the reflection b. In other words, composition of symmetries is acting like a product of sorts: $ra = b$. (By writing things in this way, we mean that transformation r is carried out first, followed by transformation a.) In fact, the composition of any two transformations is one of the six original symmetries we listed earlier.

Puzzle 34: Composing Symmetries.

Suppose you transform a triangle by the reflection a followed by the reflection b, as denoted above. To what symmetry is this product ab equivalent? Is it the same as the product ba, in which these reflections are applied in the reverse order?

We can make the following "multiplication" table that shows us precisely what happens when we compose any two symmetries. For example, row r intersects column a at entry b, representing our previous discovery that $ra = b$.

	i	r	R	a	b	c
i	i	r	R	a	b	c
r	r	R	i	b	c	a
R	R	i	r	c	a	b
a	a	c	b	i	R	r
b	b	a	c	r	i	R
c	c	b	a	R	r	i

Sudoku aficionados will notice that every row and column of the table contains each of the six symmetries i, r, R, a, b, and c exactly once. If you stare hard at the table, you might notice a few other points of interest. For one thing, each transformation can be undone by another transformation. For example, if you rotate through 120 degrees with r, and then rotate through 240 degrees with R, then the triangle is back in its starting position. Composing those two symmetries in succession is equivalent to applying the identity transformation. Likewise, any of the three reflections a, b, or c, when composed with itself, produces the identity. Another interesting property is that composition is associative. You might remember the associative property as being the one that says

$$A(BC) = (AB)C.$$

No doubt you learned that as a fact about multiplication of numbers. It turns out that if those letters represent symmetries, and the implied operation is composition, then the equation remains true.

This sort of thing seems to happen all the time in mathematics. You have a set, and some sort of operation that allows you to combine pairs of elements to get a third element. Moreover, this operation possesses certain pleasing attributes. We have just seen this for the symmetries of an equilateral triangle. The symmetries of any other geometrical object would have worked just as well.

For another example, consider the set of integers (positive, negative, and 0). We can certainly add two integers to get a third, and once we have added two numbers, we can get back where we started from by adding an appropriate negative number. On the other hand, multiplication of integers does not behave quite so nicely. In order to undo multiplication, we need to be able to divide, but that requires fractions.

There is no shortage of other examples. And when mathematicians notice that many different objects drawn from different branches of mathematics all have the same properties, they typically make an abstract definition in their honor. We shall refer to these sorts of objects as *groups*. Here is the formal definition.

Definition 1
Let S *be a nonempty set. Let* $*$ *denote a binary operation on* S *satisfying the following properties:*

(i) The operation $*$ *satisfies the associative property.*
(ii) There exists an element $e \in S$ *such that if* $x \in S$*, then* $x * e = x$ *and* $e * x = x$.
(iii) For every $x \in S$ *there exists a* $y \in S$ *such that* $x * y = e$ *and* $y * x = e$.

Then the combination of the set S *with the operation* $*$ *is referred to as a group.*

Perusing that definition gives us another opportunity to emphasize a point we have made before. Relatively straightforward ideas can be made to seem very complicated when expressed with sufficient precision.

The key to parsing such a definition is to see how it applies in a concrete example. We have done this informally in the case of the symmetries of the triangle. Now let S be the set of integers and let $*$ denote familiar addition. Then the integer playing the role of "e" is 0, because anything added to 0 gives you the number back again. Now pick any integer x. I can multiply x by the integer -1 to obtain $-x$. We then have $x + (-x) = 0$. We can therefore say that the set of integers is a group with respect to the operation of addition.

Just as with the equilateral triangle, the symmetries of a Sudoku square form a group with respect to composition. We will make use of that fact in the remainder of this chapter, and if you are really eager to complete our counting problem then you may skip ahead to the next section. There are, however, a few other points to ponder about our shiny new definition.

At some point during your education, probably in middle school, you learned the associative property and the commutative property of numbers, and you learned that they apply to both addition and multiplication. Since we have already reminded you of the associative property, let us now remind you that the commutative property (for addition) says that if x and y are two numbers, then $x + y = y + x$. If your education was like ours, then your teacher made a very big deal about such things. You probably wondered what all the fuss was about, since for numbers, these properties, let's face it, are *really* obvious.

The reason for calling attention to them comes when you start studying sets whose operations are not so well-behaved. For example, the operation of composing symmetries is not commutative. We have already seen that if we rotate a triangle through 120 degrees with the symmetry r and then reflect across the vertical axis with the symmetry a, the result is the reflection across the axis that we called b. In other words, $ra = b$. In the other order, applying a first and then r, we get $ar = c$. Therefore, our operation is not commutative. The fact that we cannot, in general, assume our operation is commutative is the reason we expressed parts (*ii*) and (*iii*) of our definition (emphasizing, for example, that both $x * e$ and $e * x$ equaled x) in the way we did.

In presenting the definition of a group, we have entered the branch of mathematics known as *Abstract Algebra*. The word *algebra* comes to us from Arabic words meaning, roughly, "the reunion of broken parts." That is why algebraic rules typically involve the proper methods for combining elements of sets with respect to given operations. The commutative and associative properties are examples of algebraic rules.

The "abstract" part refers to the idea that we no longer care what the symbols actually represent. We only care about the rules that allow us to manipulate them. Notice that the definition of a group talked only about sets and operations. No assumption at all was made regarding what the elements of the set actually are. They could be numbers, functions, matrices, other sets, or something even more exotic.

Does that mean the sort of algebra you learned in high school should be called *Concrete Algebra*? Perhaps so. It was concrete in the sense that you took it for granted that the xs and ys in the equations represented numbers. That was nice, because numbers are very familiar objects. But as you move through the higher echelons of mathematics things get far more obscure. Suddenly you are studying sets and operations that are defined very abstractly, at which point you must work out the proper rules for yourself.

There are many things you did in high school algebra that you probably never thought about too carefully. Here is an example. How would you solve this equation:

$$X^2 - 5X + 6 = 0?$$

That is a quadratic equation, of course, and you surely learned that factoring can be a very effective method. We get

$$(X - 2)(X - 3) = 0.$$

You now concluded that this is possible only if at least one of the factors is equal to 0. That happens only if $X = 2$ or $X = 3$, and that is the solution.

By using that method, you assumed that if the product of the two factors is 0, then at least one of the factors is 0 as well. And that is fine, so long as X represents a number. You have known for years that the only way to multiply two numbers together to get 0 is if one of the numbers is already 0.

It has been wisely said that you do not appreciate what you have until it is gone. Consider matrices. You might recall that matrices can be both added and multiplied, with multiplication being given by the following, somewhat complex, rule:

$$\begin{pmatrix} a & b \\ c & d \end{pmatrix} \begin{pmatrix} e & f \\ g & h \end{pmatrix} = \begin{pmatrix} ae + bg & af + bh \\ ce + dg & cf + dh \end{pmatrix}.$$

(Addition, recall, is carried out in the natural way by adding corresponding components.) If this definition is new to you, you can test your understanding by showing that multiplying any matrix by $\begin{pmatrix} 1 & 0 \\ 0 & 1 \end{pmatrix}$ will give you the original matrix back again (just like multiplying a number by 1 does not change anything.) There are good reasons for this elaborate definition, typically revealed in a course in linear algebra, but for now we will just ask you to take on faith that it is reasonable.

What if the X in our quadratic equation represents a matrix? That is possible. We would then take X^2 to mean the matrix multiplied by itself. The matrix $5X$ would be found by multiplying each entry in X by 5. The 6 would then be understood to mean 6 times the matrix $\begin{pmatrix} 1 & 0 \\ 0 & 1 \end{pmatrix}$ (just like the number 6 in our polynomial could be thought of as 6×1). The 0 on the other side then represents the matrix whose entries are all 0.

The result is a perfectly good matrix equation. We could even factor it just like we did before, because matrices satisfy the same rules of multiplication and addition as numbers. But now we have a problem. You see, it is possible to take two matrices, neither of which is 0, multiply them together and obtain 0. Here is an example:

$$\begin{pmatrix} 1 & 0 \\ 0 & 0 \end{pmatrix} \begin{pmatrix} 0 & 0 \\ 0 & 1 \end{pmatrix} = \begin{pmatrix} 0 & 0 \\ 0 & 0 \end{pmatrix}.$$

This means that the factoring approach to solving quadratic equations, which works so well for numbers, no longer works for matrices.

And *that* is why you have to think hard about the algebraic properties your letters satisfy!

5.6 BURNSIDE'S LEMMA

While it is certainly an amusing observation that the composition of two permutations is another permutation, why did we get so excited that the symmetries of the Sudoku square actually form a group?

The reason is that mathematicians have been studying groups for a long time and actually know quite a lot about them. This body of knowledge is referred to as *Group Theory*, and we can now deploy it as a major weapon in our fight against Sudoku-related ignorance.

When you are first introduced to abstract mathematics, the objects of study tend to sit lifelessly on the page. There are superprecise technical definitions and theorems that are difficult to parse, and you work very hard just to understand what everything means. It is all rather frustrating. With practice and experience, however, that starts to change. The objects come to life. They interact with one another. They *do* things.

Picture a Sudoku square in your mind. His name is S. Now imagine the group of symmetries standing off to one side of S. All of the individual symmetries are lined up, each waiting its turn to act on S. One by one they file past, each one representing a different permutation of the cells of S. So the first one goes past, permutes the cells of S, and produces a new square. Then the second one goes past, permutes the cells of our original square S, produces yet another square, and moves on. Two new squares have been obtained from S, so now there are three squares. It is possible that these squares are equivalent under a relabeling, but let that pass. Eventually all of the symmetries have done their thing, and the result is a big collection of squares, none of which are fundamentally different from S. This big collection of squares shall be referred to as the *orbit* of S.

Having worked out one orbit, we could now choose a square not contained in it and repeat the process. We could keep doing this until every Sudoku square is found in some orbit or other. As we have seen, different orbits will contain different numbers of squares, but every square must be in *some* orbit. The number of fundamentally different Sudoku squares is then equal to the number of different orbits.

As it happens, this sort of situation comes up a lot. You have a group that acts on a set in much the way we just described. That is, each element of the group affects some change on the elements of the set. (To be precise, there is a bit more involved in the notion of a *group action*, but we have no wish to burden you with another technical definition.) The problem of counting the number of orbits in a group action arises frequently, and there is a marvelous tool for solving such problems.

That tool is known as *Burnside's Lemma*. It is best understood by seeing it in action. Imagine that the edges of an equilateral triangle are colored either red or blue. Since there are three edges, each of which can be colored in two different ways, we see that there are eight colorings in all.

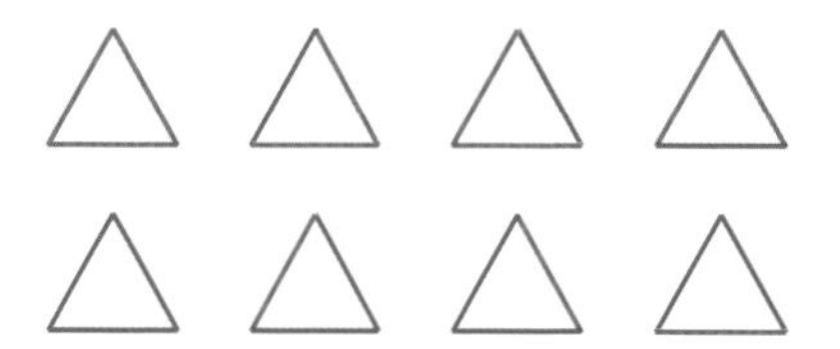

Some of these colorings are equivalent to others under the action of the symmetry group. For example, consider two different triangles, each of which has one red edge and two blue edges. Then one of the triangles can be turned into the other by applying a strategically chosen rotation. In fact, all three triangles with exactly one red edge are equivalent to one another. We can view these three triangles as representing a single orbit of the group action:

You can now see that there are actually four orbits of this action. That is, there are only four fundamentally different colorings. They can be categorized by the number of red edges they contain. You have either no red edges, one red edge, two such edges, or three:

Burnside's Lemma allows us to arrive at this answer another way. To apply the lemma we notice that certain colorings are fixed under the action of specific group elements. For example, each of these four colorings:

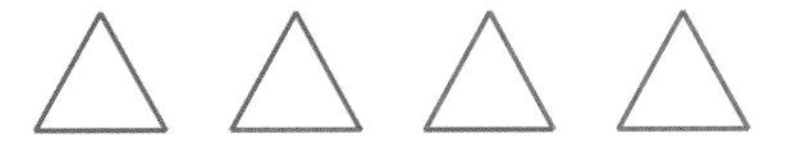

is left unchanged when we reflect across the vertical axis with the symmetry we earlier called a. Let us make a list of all six symmetries and the number of colorings they fix:

Symmetry	Fixed colorings	Number fixed
i	all eight of them	8
r		2
R		2
a		4
b		4
c		4

According to the table above, the average number of colorings fixed by the various symmetries is

$$\frac{8+2+2+4+4+4}{6} = \frac{24}{6} = 4.$$

Four also just happens to be the number of orbits we found earlier. What Burnside's Lemma says is that, surprisingly, this always works. To count the orbits in a group action, you make a complete list of all of the group elements, determine for each one the number of elements of the set they fix, and then compute the average over all of the symmetries of the number of fixed elements. The result will be the number of orbits. That the set of symmetries on the triangle forms a group that acts on the set of possible red/blue edge colorings allows us to apply the power of Burnside's Lemma to find the number of orbits. The same is true of Sudoku: Since the set of Sudoku symmetries is a group acting on the set of Sudoku squares, we can use Burnside's Lemma to find the number of orbits under the action of those symmetries, that is, the number of fundamentally different Sudoku squares.

Let's look at this in a particular Shidoku example. Consider the Sudoku symmetry where we swap the first and second columns and also swap the third and fourth columns like this:

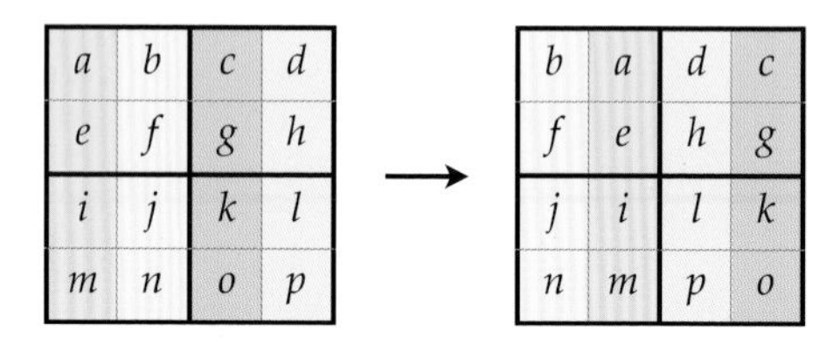

Which Shidoku squares are fixed under the action of this symmetry? Put that way, the answer obviously is that none of them are fixed. What we really need to know is how many squares are fixed *up to permutation* of the symbols. Said another way, we need to know how many Shidoku squares have the property that after performing this column-swapping symmetry there is some relabeling of symbols that will return us to the original square. Let's start by finding just one:

Puzzle 35: Burnside's Shidoku Column Swap.

Find a 4×4 Shidoku square that has the property that switching the first and second columns and switching the third and fourth columns results in a Shidoku square that differs from the first only by symbol-relabeling. To make things simpler, assume that the top-left block of the Sudoku square looks like this:

1	2		
3	4		

The solution is in the text below.

Let's walk through the process of finding a board as specified in the puzzle above. We seek a board that is just a relabeling of itself after performing our double column swap. It turns out that in this case, we know exactly what that relabeling has to be. Swapping the first and second columns completely determines the relabeling. To see this, note that whatever else we put in the board, we know that returning to the original square will require use of the relabeling $1 \leftrightarrow 2$ and $3 \leftrightarrow 4$, since that is exactly the relabeling caused by our column swapping:

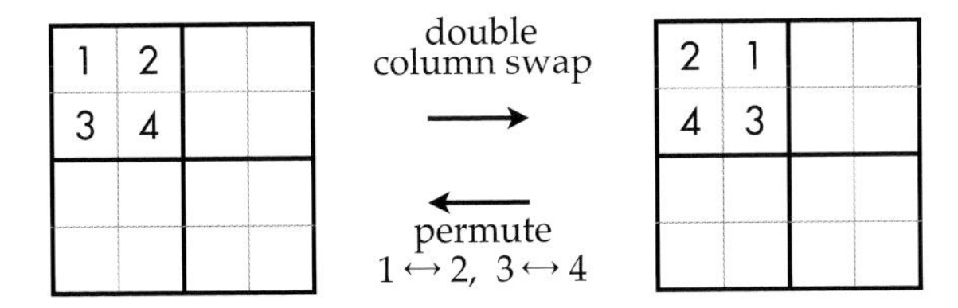

There is more than one Shidoku square that has the property we seek, so at this point we will have to make some choices. Let's try choosing 3 and 4 in order from left to right to finish the first row, and choosing 2 and 4 in order from top to bottom to finish the first column. Given these decisions, the column-swapped board, with one choice in blue and the other in green, looks like this:

1	2	3	4
3	4		
2			
4			

double column swap →

← permute $1 \leftrightarrow 2,\ 3 \leftrightarrow 4$

2	1	4	3
4	3		
	2		
	4		

Given that the two boards we are following must differ by the permutation that interchanges 1s with 2s, and 3s with 4s, we are forced to fill in the red numbers below:

1	2	3	4
3	4		
2	1		
4	3		

double column swap →

← permute $1 \leftrightarrow 2,\ 3 \leftrightarrow 4$

2	1	4	3
4	3		
1	2		
3	4		

Considering the rules of Shidoku, we now have some more forced red entries:

1	2	3	4
3	4		
2	1	4	3
4	3		

double column swap →

← permute $1 \leftrightarrow 2,\ 3 \leftrightarrow 4$

2	1	4	3
4	3		
1	2	3	4
3	4		

We have one more choice to make. After making the purple choice of 1 and 2 in order from left to right to finish the second row, we can use the swapping property and the rules of Shidoku to fill in the two boards like this:

1	2	3	4
3	4	1	2
2	1	4	3
4	3	2	1

double column swap $\longrightarrow$

$\longleftarrow$ permute $1 \leftrightarrow 2,\ 3 \leftrightarrow 4$

2	1	4	3
4	3	2	1
1	2	3	4
3	4	1	2

The board on the left above is the one we sought in Puzzle 35. If you look back through the argument above, you will see that we made three binary choices along the way, which we marked in blue, green, and purple. Therefore, there are $2 \cdot 2 \cdot 2 = 8$ possible boards with top-left square fixed in order that have the property we seek. Eight boards are fixed up to permutation by our double column-swap symmetry.

Most symmetries don't fix any Shidoku boards at all up to permutation. For example, swapping just the first and second column cannot be undone by a relabeling for any board. Others fix fewer or more than eight boards up to permutation. For example, the transpose symmetry only fixes two squares up to permutation. We'll let you do the heavy lifting on that one.

Puzzle 36: Burnside's Shidoku Transpose.

Find *two* different 4×4 Shidoku squares with the property that reflecting over the upper-left to lower-right diagonal results in a Shidoku square that differs from the first only by symbol relabeling. Assume that the top-left block of the Shidoku square is in our standard 1, 2, 3, 4 order.

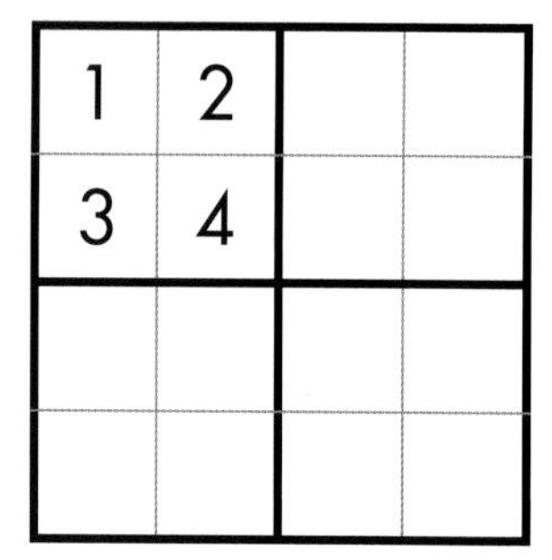

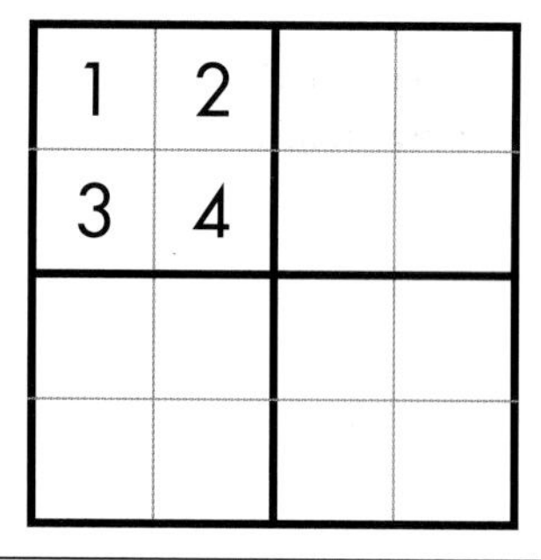

If we repeated this type of analysis for all possible Shidoku symmetries, we could find the average number of boards fixed up to permutation. It is tedious to work out all the cases by hand, but with the assistance of a computer Arnold and Lucas [5] have shown that with a group of 128 Shidoku symmetries and fixed first block, there are 56 symmetries that fix no Shidoku squares, 48 symmetries that fix 2 Shidoku squares, 9 that fix 4, 4 that fix 6, 6 that fix 8, 4 that fix 10, and only 1 symmetry (the identity) that fixes all 12 Shidoku squares up to permutation. Therefore across all symmetries, the average number of squares fixed is

$$\frac{56(0) + 48(2) + 9(4) + 4(6) + 6(8) + 4(10) + 1(12)}{128} = \frac{256}{128} = 2.$$

Burnside's Lemma now tells us this is equal to the number of orbits of Shidoku squares. In other words, that there are exactly two fundamentally different Shidoku squares.

That was a lot of work to find out something we already knew! However, we can take these same methods to the 9×9 Sudoku case to find out something that we don't know already: the number of fundamentally different Sudoku squares.

5.7 BRINGING IT HOME

Alas, now we must once more bring in the computer and wave our hands.

Determining the precise size of the Sudoku symmetry group is a difficult problem. Keep in mind that the basic permutations on our list can be combined into long chains, and it is very difficult to determine which chains produce new permutations and which are repeats of ones we have already seen. Suffice it to say that it is a very large group indeed. The square S will be waiting for quite some time for all of the symmetries to file past. The number is too large to permit a direct application of Burnside's Lemma.

Russell and Jarvis [36], were able to circumvent this problem by taking advantage of more subtle considerations than we have discussed here. They were able to divide the vast number of symmetries into 275 classes with the property that any 2 symmetries in the same class had the same number of fixed squares. *That* is a number with which one can deal.

Moreover, it is easy to see that many symmetries will have no fixed squares at all. For example, consider the reflection across the vertical axis passing through the middle of the fifth column. Below left we see a partially filled-in Sudoku square, and below right is its reflection across this center vertical axis:

A	B	C	D	4				
				6				
				2				
				9				
				5				
				1				
				7				
				8				
				3				

				4	D	C	B	A
				6				
				2				
				9				
				5				
				1				
				7				
				8				
				3				

Suppose this square was fixed, up to relabeling, by the center reflection symmetry. Since the reflection leaves the fifth column completely unchanged, the relabeling in question is forced to leave all the digits as they are. This means that the reflected square would have to be the very same square that we started with. But that is impossible, since it implies that the entries A, B, C, and D shown in the first row actually appear twice each. The center vertical reflection symmetry cannot fix this or any Sudoku squares.

Taking advantage of these properties allowed Russell and Jarvis [36] to apply Burnside's Lemma to this problem. They found that the total number of fundamentally different Sudoku squares is 5,472,730,538. A large number, but quite a comedown from the 10^{21} many squares we found in the last chapter.

A natural question about Sudoku puzzles has once more led us, inexorably and unavoidably, to major ideas in higher mathematics. We find that satisfying.

So what does this mean about *playing* Sudoku? We have already seen that different puzzles, with different difficulties, can have the same solution square. Likewise, puzzles whose solutions are equivalent Sudoku squares can nonetheless be very different puzzles. What would it take for two Sudoku *puzzles* to be essentially equivalent? For that to happen, they would have to differ from each other by a sequence of Sudoku symmetries and relabeling of symbols. If two puzzles are related by those symmetries and relabelings, then those two puzzles should be exactly the same level of difficulty and in fact require exactly the same sort of solving techniques.

Puzzle 37 and 38: Sudoku Clones.

Fill in each grid so that each row, column, and block contains each of the number 1–9 exactly once. These puzzles are fundamentally equivalent in the sense that you can get from one to the other by a sequence of Sudoku symmetries and relabelings. Can you tell while playing that the puzzles are essentially the same? (Watch out; depending on the solving techniques you are familiar with, you may need to use Ariadne's Thread at some point.)

	6					2	1	
				9	2			
		2	1	8				3
	5	4			8			9
	9						8	
2			4			6	3	
4				5	3	1		
			7	2				
	7	1					9	

6				7	3	2		
			5	1				
	5	2					9	
1			6			4	3	
	9						8	
	7	6			8			9
	4					1	2	
				9	1			
		1	2	8				3

Bonus Round: Find a sequence of relabelings and Sudoku symmetries that carries the first puzzle to the second.

6

Searching

The Art of Finding Needles in Haystacks

There are a great many partially filled-in 9×9 squares, but most of them are not sound Sudoku puzzles. Some of them may have no completions, because the starting clues are contradictory. Others may have thousands of completions. Still others might have a single solution, but the puzzle is so trivial that it is not enjoyable to solve. How do we find the few grains of wheat amid the endless vistas of chaff?

Since making a 9×9 Sudoku square by hand can be a time-consuming process, we shall work instead with the 6×6 squares called *Roku-Doku*. Here are two examples:

Puzzles 39 and 40: Roku-Doku.

Complete each grid so that each row, each column, and each of the six 2×3 blocks contain each of the numbers 1–6 exactly once.

		3	2		
		4			
	1				5
5				6	
			3		
		1	5		

	6			5	
	3				2
		4			
			1		
2				1	
	4			6	

6.1 THE SUDOKU STORK

In principle, there are many ways to go about making a puzzle. The simplest approach begins with a completed square and systematically removes entries until an acceptable puzzle remains. Let's use this method to construct a Roku-Doku puzzle whose solution is the completed square shown on the left in the following diagram. Technically this square is a puzzle, since it has a unique solution. But it seems rather insipid, since there are no entries to fill in! Let's start our quest for a puzzle by removing the entries in the four corners of the square, as shown below right. Since it is generally considered aesthetically pleasing for the starting clues to be placed in a rotationally symmetric pattern around the center of the square, we shall always remove cells in symmetric pairs.

2	3	6	1	5	4
4	1	5	2	3	6
1	2	3	4	6	5
5	6	4	3	2	1
6	4	2	5	1	3
3	5	1	6	4	2

	3	6	1	5	
4	1	5	2	3	6
1	2	3	4	6	5
5	6	4	3	2	1
6	4	2	5	1	3
	5	1	6	4	

Why start by removing the four corners? No particular reason. It just gives the puzzle a pleasant, rounded look.

Our cornerless Roku-Doku clearly has a unique solution. Let us now remove some more cells. You might think that, with so many filled-in cells, there would be little danger of rendering the puzzle unsound just by vacating a few more. But care is needed even at this early stage! For example, what happens if we remove the four entries in the center of the board, as shown below left? The puzzle no longer has a unique solution. We could simply reverse the 3s and 4s in the central squares and the result is a second, correct solution to our puzzle.

	3	6	1	5	
4	1	5	2	3	6
1	2			6	5
5	6			2	1
6	4	2	5	1	3
	5	1	6	4	

	3	6	1	5	
4	1			3	6
1	2	3	4	6	5
5	6	4	3	2	1
6	4			1	3
	5	1	6	4	

On the other hand, we could remove the 2s and 5s from the central columns to obtain the puzzle above right. To see that this is sound, pretend that you were given this puzzle in a magazine and try to solve it. You will find that each of the vacant cells can have only one value.

We could continue removing cells randomly, but there are other aesthetic considerations. It would not be very appealing to have our final puzzle contain an already completed Sudoku region or to contain a region with only one vacant cell. We also would not want our puzzle to contain every representative of a particular digit. It would be rather boring if our puzzle already contained six 1s, for example.

With those considerations in mind, deleting the 1s and 6s from rows 3 and 4 seems like a good move, leaving the puzzle below left. Since the second and fifth columns still have five representatives you might try removing their 2s, but this would be a mistake. The pattern of 2s and 6s in rows 3 and 4 is much like the pattern of 3s and 4s in those rows. They can be swapped to produce a new, equally valid solution. On the other hand, removing the 3s from rows 3 and 4 does leave a sound puzzle. It is on the right below:

	3	6	1	5	
4	1			3	6
	2	3	4		5
5		4	3	2	
6	4			1	3
	5	1	6	4	

	3	6	1	5	
4	1			3	6
	2		4		5
5		4		2	
6	4			1	3
	5	1	6	4	

We still have five 4s, while every other digit is represented no more than four times. With a bit of experimentation, you might notice that the 4s in columns 2 and 5 can be removed, along with the 3s in those columns (to preserve symmetry), without affecting the puzzle's soundness. The result is the puzzle below left.

Things are going well. Our puzzle satisfies all of our aesthetic requirements. It is rather easy to solve, however. To remedy that we might wish to vacate a few more cells, keeping in mind that we want to remove digits in symmetrical pairs.

Trial and error reveals that the 1s and 6s in the first two and last two columns can be removed. But that is the last of the redundant pairs. The finished puzzle is below right.

		6	1	5	
4	1				6
	2		4		5
5		4		2	
6				1	3
	5	1	6		

		6	1	5	
4					
	2		4		5
5		4		2	
					3
	5	1	6		

Not too shabby for our first time out. However, if you play this puzzle you will find that it is fairly simple. If we had chosen different entries to remove, then we may have produced a more difficult puzzle. But how do we know which pairs to remove?

6.2 A STORK WITH GPS

To this point, we have mostly been messing around. There is no shame in that; trial and error is an honorable approach to solving mathematical problems. Still, we worked awfully hard to produce a puzzle that was not entirely satisfactory. Perhaps it is not such a bad idea to look into more sophisticated approaches.

Let us try the opposite direction, by starting with a blank board and *adding* pairs of entries. Our target will be the solution square from the previous section. We shall place clues in symmetric pairs until an acceptable puzzle emerges. But how are we to begin?

Look at the four orange cells on our solution board shown below left. These four cells form an *unavoidable set* in the sense that every puzzle with this square as a solution must have at least one clue in this set. The reason is the toggling phenomenon we saw before. Were every cell save for these four filled in, we still would have two valid solutions (by switching the 2s and 6s in these cells). The purple cells below right are likewise an unavoidable set for this particular Roku-Doku square.

2	3	6	1	5	4
4	1	5	2	3	6
1	2	3	4	6	5
5	6	4	3	2	1
6	4	2	5	1	3
3	5	1	6	4	2

2	3	6	1	5	4
4	1	5	2	3	6
1	2	3	4	6	5
5	6	4	3	2	1
6	4	2	5	1	3
3	5	1	6	4	2

Scanning our solution square reveals five other unavoidable sets, for a total of seven. They are highlighted below left. In filling in our blank board, we can begin by selecting, as efficiently as possible, a group of cells containing at least one representative from each set. If we maintain our insistence on symmetrical pairs, it turns out we need ten initial clues. There are many ways of selecting those clues, one of which is shown below right.

2	3	6	1	5	4
4	1	5	2	3	6
1	2	3	4	6	5
5	6	4	3	2	1
6	4	2	5	1	3
3	5	1	6	4	2

		5			6
1	2		4		
		4		2	1
6			5		

A careful scan reveals that no other cells are determined by these ten clues, meaning we do not yet have a sound puzzle. We must add more starting clues, but we want to do this as efficiently as possible. That is, we want to be certain that our additional clues are as informative as possible. Toward that end, consider the diagram below left. The red digits between neighboring cells represent values that must be placed into one of those cells. This suggests that additional clues placed among those cells will be quite helpful. It turns out that adding clues in the upper-right and lower-left corners is sufficient to produce a sound puzzle, as shown below right.

		6			5
		5			6
1	2		4		5
	5	4		2	1
6			5		
	5			6	

					4
		5			6
1	2		4		
		4		2	1
6			5		
3					

This was a considerable improvement over the last section. A few simple considerations helped us construct a more challenging twelve-clue Roku-Doku puzzle.

Puzzles 41 and 42: Roku-Doku Redux.

Complete each grid so that each row, each column, and each of the six 2×3 blocks contains each of the numbers 1–6 exactly once. These are the two puzzles we constructed above; both have the same solution that we started with.

		6	1	5	
4					
	2		4		5
5		4		2	
					3
	5	1	6		

					4
		5			6
1	2		4		
		4		2	1
6			5		
3					

6.3 HOW TO SEARCH

Everything we have done is readily automated. A computer can be programmed to produce valid Sudoku squares, and then, either by removing symmetrical pairs or by placing symmetrical pairs, produce a sound puzzle. Given a partially filled square, the computer can easily determine the number of valid completions. The difficulty lies not in producing sound puzzles, but in finding those that are interesting, challenging, and beautiful.

The quest for sound Sudoku puzzles is an example of a search problem. Such problems have the following general form: We have a large collection of objects.

Each of these objects has a score attached to it measuring its desirability for our purposes. We desire objects possessing high scores. But the collection is too large to examine each object individually. We can examine only a small percentage of the total space. How should we conduct our search to maximize our chances of finding the desired objects?

From now on we shall refer to the "large collection of objects" as the *search space*. It can be helpful to imagine your search space as a surface with peaks and valleys. The peaks represent high-scoring points and the valleys represent low-scoring points. Most points lie somewhere between these extremes. The surface in this metaphor is commonly referred to as *the search landscape*. In terms of Sudoku, the search landscape contains all partially filled-in Sudoku squares, and desirability is determined by the number of completions the pseudo-puzzles admit. The goal is to locate a partially filled-in Sudoku square that admits just one solution.

The simplest of all approaches is *random search*. We select our objects at random, and we get what we get. It is a primitive method, but often effective. Specifically, it will be effective when the number of desirable objects in the space is a relatively large percentage of the total. For example, suppose just one-tenth of 1 percent of the objects have the properties we seek. If we choose randomly, then we would expect that one out of every one thousand objects will be desirable. Using a computer, we can typically search many millions of possibilities in a short period of time. In such a situation, random search works rather well.

The number of partially filled-in Sudoku squares is enormous, but with a fast computer we could likely do a random search, choosing subsets of a Sudoku square until one is found with a unique solution. We could also do a more direct search, removing pairs sequentially until no more can be removed, as we did at the start of this chapter. It does not take long at all to produce a sound puzzle in this way.

The problem comes when our criterion is something more than mere soundness. We might seek to minimize the number of starting clues, for example. For Sudoku puzzles, it is currently thought that eighteen is the minimum number of clues for a sound, symmetrical puzzle. (For a nonsymmetrical puzzle, seventeen clues appears to be the minimum.) Beginning with a valid solution square and selecting eighteen cells at random is unlikely to produce a sound puzzle. It is a straightforward calculation to show that the number of ways of selecting eighteen cells from a total of eighty-one is on the order of 10^{16}, which is too large for an exhaustive search. Finding such puzzles in a reasonable amount of time requires a better method.

If random search is impractical, what are the alternatives? One possibility is known as a *hill-climbing algorithm*. We pick our initial point at random, but from there we confine our search to the local area around that point. We look for nearby points possessing a higher score than our current point. If we find one, then we use this second point as the beginning of a new, local search. This continues until we can no longer find a nearby, higher-scoring point. As the name suggests, this works rather well if your landscape looks like a simple pile of sand. Your initial, randomly chosen point may have a low score near the bottom of the pile. But no matter where you begin, you will always work your way to the top. The reverse situation works as well. If your landscape resembles a bowl and you wish to find the lowest point, you can with confidence employ a *hill-descending* algorithm.

On the other hand, these algorithms perform poorly when the landscape has many peaks and valleys. You will certainly climb to the peak of whichever hill you chance upon, but you are unlikely to find the globally highest point. Worse, you will have no way of knowing whether your particular peak represents a local or a global peak. Hill-climbers also fare poorly when the landscape is highly discontinuous. If the landscape is very jagged, so that high points and low points are thoroughly mixed together, then you will need a different sort of algorithm.

We could also try the so-called *evolutionary algorithms*. The idea is to mimic the processes of genetic mutation and natural selection lying at the heart of the evolutionary process in nature. Our search begins by choosing several points at random. We examine those points and select a subset consisting of those which, by chance, have the highest score. These are allowed to vary randomly, producing a new set of "offspring" points. We then repeat the process. Such algorithms have proven themselves effective in solving a variety of engineering problems. On the other hand, like the hill-climbers, they are only effective when the landscape is tolerably smooth and regular.

What we *really* need is an all-purpose algorithm, one that will be effective regardless of the landscape to be searched. Such an algorithm would be a marvelous thing. Alas, it does not exist. If an algorithm performs especially well on one kind of landscape, then I promise you there are other landscapes on which it works very poorly. More precisely, the average performance of a given search algorithm over all possible landscapes is no better than a random search. This is a consequence of a collection of results known collectively, and somewhat facetiously, as the *No Free Lunch Theorems* [47].

So sad. Solving a search problem requires tailoring the algorithm to the landscape. In math, as in life, there is no substitute for hard work. And it takes a lot of hard work, and a mix of human and computer techniques, to find an eighteen-clue needle in the Sudoku haystack.

6.4 SEARCHING FOR EIGHTEEN-CLUE SUDOKU

How would we go about constructing a sound Sudoku puzzle with eighteen symmetrical starting clues? It just so happens that one of the authors of this book is also the coauthor of an anthology of such puzzles and can say from experience that such puzzles can be very difficult to find. The methods used to produce rotationally symmetric eighteen-clue puzzles are illustrative of the sort of enlightened trial and error so often required in solving difficult search problems. The process we shall describe is a condensed version of the actual process used to make the puzzles in Riley and Taalman [33].

The trouble is that we know almost nothing regarding the shape of the landscape. Are valid puzzles located close to each other in this space? If we have an initial placement of clues permitting ten completions, will small changes to that placement bring us to pseudo-puzzles with eleven or nine completions? Or do we have instead a situation in which the nearest neighbors to the ten-completion pseudo-puzzles have many thousands of solutions?

It is unclear how to get a mathematical grip on these questions. In such situations, we can only experiment and see what happens.

As a starting point, let us select a symmetric pattern of eighteen cells to use for our initial clues. We will refer to this pattern as the *mask* of the puzzle. In practice, very few masks will be suitable for finding eighteen-clue puzzles in a feasible period of time. In fact, many masks can never form valid puzzles. It takes time, experience, and luck to recognize which masks are likely to work. Here is an example of a promising-looking mask—visually attractive, not too clumpy, and not too spread out:

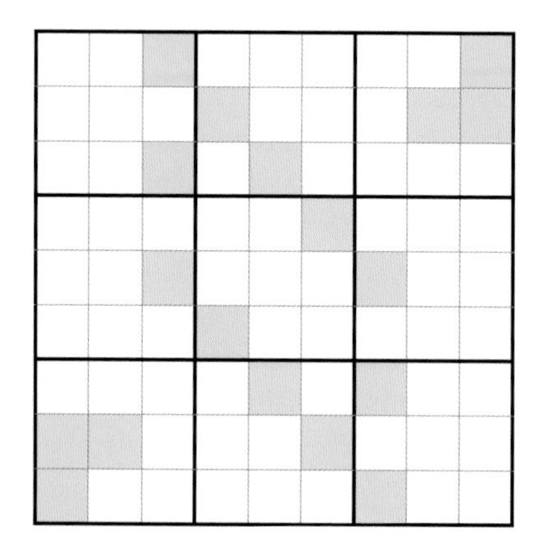

Now imagine populating this mask with specific digits, being careful not to violate any Sudoku rules. The resulting pseudo-puzzle could have many completions, and some ways of populating the mask will be better in this regard than others. Note, for example, that the starting clues in a sound puzzle must include representatives of at least eight of the nine digits. If we initially omit all instances of two digits, then these digits could be switched in any completion of the square to produce a second solution. Furthermore, we must be careful not to put too many representatives of any one digit into our mask, lest we have insufficient space for information regarding the other digits.

It is generally easier to find sound puzzles starting from a mask with a relatively even distribution of digits. Let us say we populate the mask with digits so that one number appears exactly three times, one number appears just once, with all others appearing twice. For example, we could fill up the the mask with three 2s; one 4; and two each of 1, 3, 5, 6, 7, 8, and 9, placed in some non-contradictory fashion, as shown below left:

Pseudo-puzzle with 23,593 solutions

		5						2
			7				6	3
		9		8				
					1			
		2				3		
			2					
				4		8		
6	7				5			
9						1		

Pseudo-puzzle with 18,111 solutions

		5						2
			7				6	3
		9		8				
					1			
		2				3		
			5					
				4		8		
6	7				2			
9						1		

This is just a starting point. It would be nice if it were a valid eighteen-clue puzzle, but it is not. However, 23,593 solutions is not really so many; it could have been far worse. The hope is that this grid is sufficiently close to some valid eighteen-clue puzzle that small alterations will produce what we want. Notice, for instance, that swapping the entries in the two highlighted cells shown above right results in a pseudo-puzzle with just 18,111 solutions.

This suggests an initial strategy of *reduction* by swapping random pairs. This will leave unchanged the distribution of the numbers 1–9. If swapping a particular pair of cells results in an invalid pseudo-puzzle or does not reduce the number of solutions, then we will not perform the swap. If swapping a pair of cells results in a valid pseudo-puzzle with fewer than 23,593 solutions, then we will carry out the swap, just as we did above.

This procedure will not necessarily produce a puzzle with a unique solution, and the order in which we perform the swaps might affect our ability to improve a given pseudo-puzzle. Imagine that the individual cells are numbered as shown below left. One possible search path, found with the help of a computer, involves swapping cells 49 and 69 (as shown earlier), then cells 21 and 39, then 18 and 59, then 39 and 43, then 3 and 79, then 13 and 43, and finally cells 17 and 18. Each swap reduces the number of solutions, leaving us with the 42-solution pseudo-puzzle shown below right:

Numbering the cells on the board

1	2	3	4	5	6	7	8	9
10	11	12	13	14	15	16	17	18
19	20	21	22	23	24	25	26	27
28	29	30	31	32	33	34	35	36
37	38	39	40	41	42	43	44	45
46	47	48	49	50	51	52	53	54
55	56	57	58	59	60	61	62	63
64	65	66	67	68	69	70	71	72
73	74	75	76	77	78	79	80	81

Reduction to a 42-solution pseudo-puzzle

		1						2
			9				4	6
		2		8				
					1			
		3				7		
			5					
				3		8		
6	7				2			
9						5		

Unfortunately, there are no more simple pairs of swaps that will further reduce the number of solutions. We have followed our reduction path as far as it will go, but have not yet found our needle. Surely, though, we feel *close.*

What next? We shall try a new strategy. By a *mutation,* we mean a swap of two cells that does not reduce the number of solutions. The hope is that the mutated pseudo-puzzle has a reduction superior to our original. For example, suppose we swap cells 9 and 73 to get the mutated pseudo-puzzle below left. This has far more than forty-two solutions, but perhaps it is close to a sound puzzle. It is, at least, close to our 42-solution pseudo-puzzle. By once more applying our reduction strategy we can whittle things down to the pseudo-puzzle shown below right, with just thirty-one solutions. (Can you tell what swaps happened?)

Mutation has 9,403 solutions

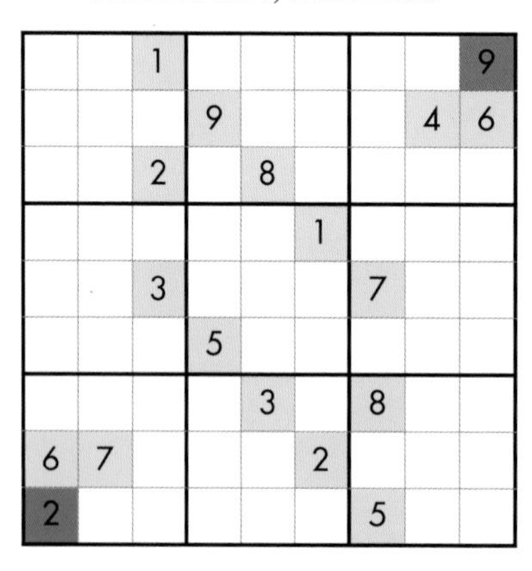

Reduce to a 31-solution pseudo-puzzle

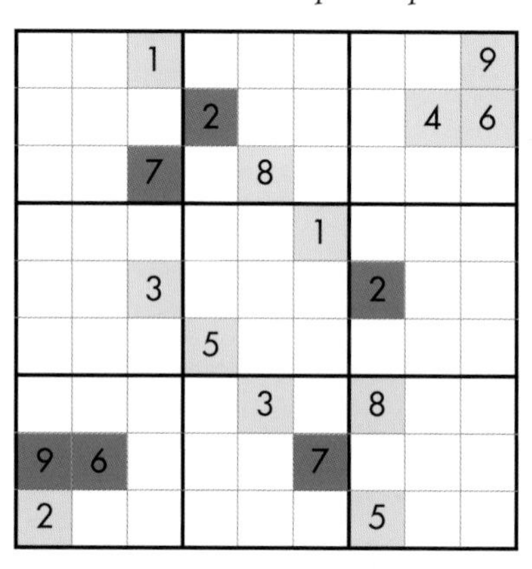

Most mutations, alas, do not improve our situation. But each time mutation-followed-by-reduction produces a superior pseudo-puzzle, we will further mutate and reduce from there. Of course, we might get unlucky, producing a situation where none of the $\binom{18}{2} = 153$ possible mutations admits a useful reduction. If that happens, we follow the strategy of *giving up*. In other words, we call it a day and start over at the beginning, repopulating the eighteen cells of the mask with new numbers. A computer can execute this algorithm many times, very quickly, and if the stars align, we will find our needle.

Happily, the grid we have been following comes through in fine style. After a few dead ends, we found one that eventually reduces to just nine solutions:

Mutation has 5,596 solutions

		1						3
			2				4	6
		7		8				
					1			
		3				2		
			5					
				9		8		
9	6				7			
2						5		

Reduce to a 9-solution pseudo-puzzle

		6						2
			5				3	7
		4		1				
					6			
		8				1		
			2					
				9		8		
9	2				7			
3						5		

Many more dead ends and a bit of computing time later, we found a useful mutation of the nine-solution pseudo-puzzle:

Mutation has 3,035 solutions

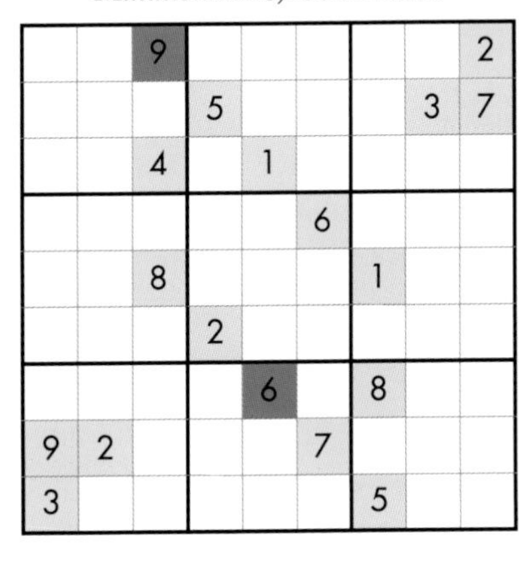

		9						2
			5				3	7
		4		1				
					6			
		8				1		
			2					
				6		8		
9	2				7			
3						5		

Reduce to a 3-solution pseudo-puzzle

		1						2
			8				5	7
		6		3				
					5			
		3				1		
			2					
				4		6		
7	2				9			
8						9		

Luckily, we can improve this as well. The mutation below left requires just a simple reduction to move us still further, this time to just two solutions:

Mutation has 1,401 solutions

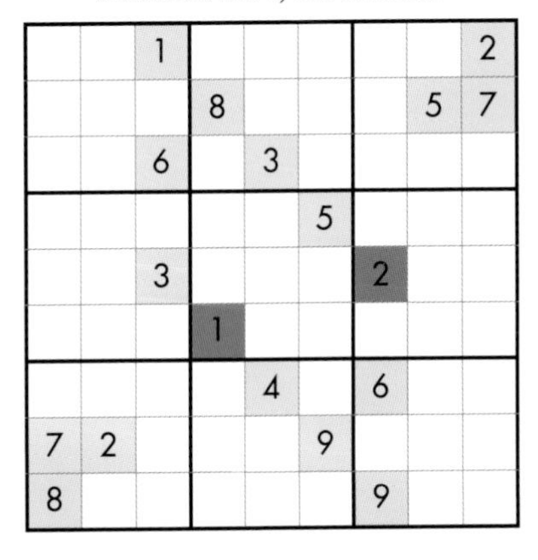

		1						2
			8				5	7
		6		3				
					5			
		3				2		
			1					
				4		6		
7	2				9			
8						9		

Reduce to a 2-solution pseudo-puzzle

		1						2
			8				5	7
		6		9				
					4			
		3				1		
			2					
				5		3		
7	2				9			
8						6		

Almost there! You should keep in mind that we are showing you the high points of this process. Many is the time our algorithm resulted in a true dead end, with no possible mutations moving us closer to a unique-solution puzzle. In some cases, even after getting this far, we found we could go no further. However, in this case, we have success!

Mutation has 226 solutions

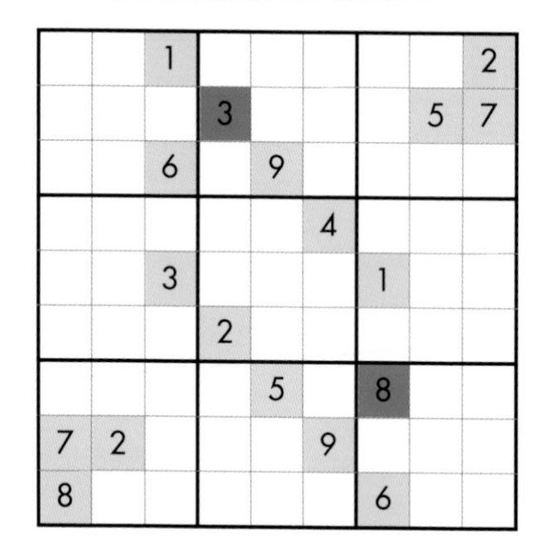

Reduce to a unique-solution puzzle!

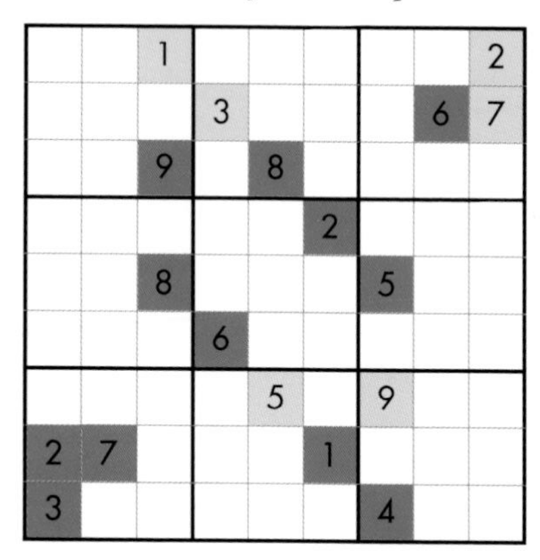

If we could have found an eighteen-clue puzzle just by a simple blind search, we certainly would have. But since sound, rotationally symmetric, puzzles are so sparse in the landscape, we had no choice but to employ a combination of randomness, directed reduction, trial-and-error, cleverness, and luck.

After all that hard work we deserve a break. Let us solve our elusive puzzle!

Puzzle 43: The Eighteen-Clue Needle.

Fill in the grid so that every row, column, and block contains each of the numbers 1–9 exactly once.

		1						2
			3				6	7
		9		8				
					2			
		8				5		
			6					
				5		9		
2	7				1			
3						4		

It turns out that once you find a mask and initial clue placement leading to success in the reduce-mutate-reduce algorithm, that same mask tends to lead to success with other initial placements. Just for fun, here are two more eighteen-clue puzzles built from the same mask:

Puzzles 44 and 45: Two More Needles.

Fill in each grid so that every row, column, and block contains each of the numbers 1–9 exactly once. Although both puzzles have the same mask as the previous puzzle, they differ by more than just permutation of symbols and are thus not equivalent as puzzles.

		4						2
			5				7	8
		6		1				
					4			
		5				1		
			9					
				6		3		
7	2				3			
9						5		

		3						8
			2				9	7
		4		1				
					8			
		1				3		
			6					
				5		4		
7	5				3			
9						6		

Throughout our search, the even distribution of the numbers 1–9 in the mask was carefully preserved at every step. However, the precise distribution we used (three of one clue, one of another, and two of all the rest) is not essential to finding an eighteen-clue puzzle. The distribution is important, but it can be relaxed. For example, here is an eighteen-clue puzzle with a different mask in which 3 appears four times; 5 and 9 each appear three times; 1 and 2 each appear twice; and the remaining numbers 4, 6, 7, and 8 each only appear once. Why that particular distribution? Because those are the first eighteen digits of $\pi = 3.14159265358979323\ldots$, of course.

Puzzle 46: Eighteen-Clue Pi.

Fill in the grid so that every row, column, and block contains each of the numbers 1–9 exactly once.

7	2							
	5				9			
				3	8			
			4			5		
		3				9		
		1			3			
			2	5				
			6				3	
							1	9

6.5 MEASURING DIFFICULTY

Let us turn to a new question. How can we assess the difficulty of a Sudoku puzzle?

The number of initial clues is one indicator of difficulty, since fewer clues provide less starting information. Generally speaking, a twenty-clue puzzle will be harder than a twenty-eight-clue puzzle. However, this general rule of thumb is not hard and fast, as is shown by the next two puzzles.

Puzzles 47 and 48: Easy Twenty and Hard Twenty-Eight.

Fill in each grid so that every row, column, and block contains each of the numbers 1–9 exactly once. The first puzzle has only twenty clues but its difficulty is a breezy level 1. The second puzzle has twenty-eight clues but its difficulty is a diabolical level 5.

	6	7						4
		2			3			
				9	8	3		
	1				5			
			9				7	
		4	2	5				
			7			1		
1						6	8	

	4	3						1
6	2			1	8		3	
1			3				5	
			5			2		
		9				1		
		2			3			
	3				1			8
	1		2	6			4	5
2						6	1	

Perhaps, then, it is the configuration, what we have been calling the "mask," of the clues that distinguishes an easy from a difficult Sudoku puzzle. Sometimes, yes. For example, one could certainly put clues into a puzzle in such a way that they provide redundant information, thereby giving less insight into the solution than you might expect. The mask of the clues has some impact on difficulty. But it is possible to make a difficult puzzle that shares the same mask as an easy puzzle, as shown below. Notice that the second puzzle is more than just a relabeling of the first; the actual distribution of the clues is different within the mask.

Puzzles 49 and 50: Easy and Hard Twins.

Fill in each grid so that every row, column, and block contains each of the numbers 1–9 exactly once. Both puzzles have the same configuration of clues. But the first is only level 1, while the second will knock you out at level 5.

	3					4		
		6			8			5
7		2				1	8	
	1			5				
			3	7	2			
				1			3	
	4	7				5		3
6			5			2		
		9					4	

	8					5		
		6			4			9
7		3				4	2	
	2			3				
			8	9	1			
				2			7	
	9	2				6		3
4			6			8		
		5					4	

While the number and configuration of clues are easy to identify, they are unreliable for characterizing the difficulty of a puzzle. A superior measure is provided by the techniques needed to solve a puzzle. As we saw in Chapter 1, some solving methods are easier to apply than others, and we might say a puzzle is difficult if complex reasoning is required. This measure has some obvious difficulties. We cannot tell just by looking at a puzzle the techniques it requires. Worse, different people might solve a puzzle in different ways or might be more comfortable with certain techniques than others. We are heading into subjective territory here.

We can nonetheless put a rough difficulty rating on the basic solving techniques. The simplest of all is the forced cell. If a given cell can contain only one value, or if a particular digit can only reside in one cell within a given region, then it is easy to progress toward a solution.

Twins are harder to rate. Finding them requires an assiduous recording of the candidate values for each cell, followed by the realization that two cells in the same region had the same candidate values. That is considerably more mental dexterity than was called for when filling in forced cells. Triples would then be more difficult still. A puzzle only requiring the identification of forced cells should be considered easier than one calling for twins or triples.

Then we come to methods like *X*-Wings and Swordfish. These require not only noticing similarities in the candidate values of multiple cells, but also some geometrical acumen. The most difficult of the methods we considered in the first chapter was that of Ariadne's Thread. We used this technique as a last resort. It involved – just writing this is painful – outright guessing. Not just that, mind you, but also then keeping track of where we made our guesses so that we could later correct them.

One way for a puzzle constructor to determine the difficulty of a puzzle is by assigning numerical scores to each technique, with more difficult techniques receiving higher scores. A computer solver can keep track of the techniques it uses, eventually leading to a numerical score being assigned to the puzzle as a whole. The higher the score, the more difficult the puzzle. A puzzle that can be solved entirely or almost entirely via forced cells, perhaps with the occasional twin, would be considered easy. If you never need Ariadne's Thread but do need the odd X-Wing or triple, then you are up to medium. Bring in Ariadne's Thread or other sophisticated solving techniques and your puzzle is hard. If long chains of deductions and techniques are needed, your puzzle is rated Fiendish or beyond.

If we want to rate difficulty without referring to particular techniques with which given human solvers may be unfamiliar, we can make a computer solver play a given puzzle thousands of times with thousands of different solution paths. Difficulty level is then assigned based on how long it takes the program to carry things out. Sometimes averaging over a large number of random solving paths is more accurate than a precise calculation based on a predetermined solving path. For puzzles whose difficulty is in question, however, nothing beats a team of human solvers who can rate the puzzles by hand.

Difficulty rating is not an exact science. The list of techniques we considered in Chapter 1 is hardly exhaustive, and some judgment calls must be made in assigning numerical scores. There may be more than one way of proceeding in a given puzzle. Perhaps some people find X-Wings difficult but Ariadne's Thread a piece of cake.

In the end, there will always be some puzzles about which people disagree. But let us be serious. We are not processing nuclear fuel here or calibrating a pacemaker. Absolute precision is not necessary.

6.6 EASE AND INTEREST ARE INVERSELY CORRELATED

All this raises another question: Why would anyone bother with a Hard or Fiendish puzzle? The Easy puzzles can be polished off without much difficulty. Why not just stick to those?

Perhaps the answer is clear. There is no satisfaction in solving an easy puzzle. If all of the cells are forced then we are reduced to the level of machines. We are just mindlessly carrying out algorithms which is, after all, what computers were invented to do. It may not be easy to spot an X-Wing or to play out Ariadne's Thread until some insight is gained, but that is why successfully applying such things brings a smile to your face. No pain, no gain, as the saying goes.

Your authors have been teaching mathematics for quite some time now, and it has been our persistent experience that this lesson, obvious when pondering Sudoku puzzles, seems to elude students struggling for the first time with calculus or linear algebra. When presented with a problem requiring only a mechanical, algorithmic solution, the average student will dutifully carry out the necessary steps with grim determination. Present instead a problem requiring an element of imagination or one where it is unclear how to proceed, and you must brace yourself for the inevitable wailing and rending of garments.

This reaction is not hard to understand. In a classroom setting there are unpleasant things like grades and examinations to consider, not to mention the annoying realities of courses taken only to fulfill a requirement. Students have more on their minds than the sheer joy of problem solving. They welcome the mechanical problems precisely because it is clear what is expected of them. A period of frustrated confusion can be amusing when working on a puzzle, because there is no price to be paid for failing to solve it. The same experience in a math class carries with it fears of poor grades.

Still, it is an interesting difference in perspective between those outside and those inside the mathematical community. For mathematicians, a problem is often considered solved as soon as an algorithm for producing a solution is found. That a clearly defined sequence of steps will lead to a solution is sufficient; the actual answer tends to be anticlimactic. There is nothing in the world more boring than an arithmetic problem. Most of us do not even see a question like, "What is $3{,}456 \times 2{,}874$?" as a math problem at all. It certainly is not the sort of thing we dwell on in our day-to-day lives.

We prefer questions like these:

Puzzle 51: Checkerboard Dominos.

Suppose the diagonally opposite corners of a standard 8×8 checkerboard are removed. Is it possible to cover the remaining squares with dominoes, where each domino covers two adjacent squares?

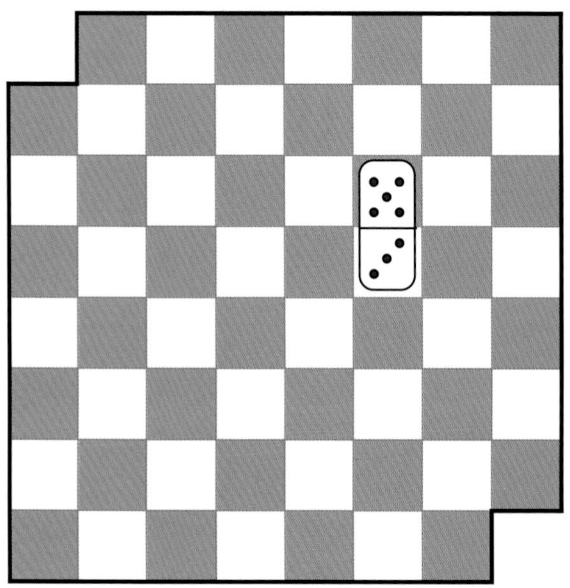

Puzzle 52: Chessboard Knights.

A knight in chess moves two squares in one direction and then one square in a perpendicular direction. Two knights are said to attack each other if one can move to the other's square. How many knights can you place on an empty 8×8 chessboard so that no two knights attack each other?

The puzzles above show the abstract type of game that mathematicians enjoy, and the importance and elegance of the "aha!" insight. In every branch of mathematics, there are certain big problems that animate the field. These are the ones everyone is trying to solve. They are intriguing precisely because so many others have tried and failed to solve them before you. In Sudoku this problem is the question of whether seventeen is truly the minimum number of clues that a sound Sudoku puzzle can have (or eighteen, if the clues are required to be rotationally symmetric). Many have tried to solve it, but so far the answer remains a mystery.

It is rare for such problems to be solved in one great flash of insight. Instead what happens is that many mathematicians chip away at the problem over periods of years or even decades. If solving the problem is too hard, you at least try to leave the problem a bit more solved than you found it. Eventually so much has been chipped away that the problem is ready to fall.

Which brings us back to the title of this section. The difficulty of a problem is inversely correlated with the interest it holds for mathematicians. Easy problems hold no interest. Hard problems are irresistible, if often frustrating.

6.7 SUDOKU WITH AN EXTRA SOMETHING

Enough talk! Time for puzzles! But by now we are old hats at traditional Sudoku, so let us up the ante.

Just as playing Sudoku variations requires different techniques from standard Sudoku, creating such variation puzzles presents its own set of challenges. Think back to the Four Square Sudoku from Puzzle 7, where we had four additional 3×3 blocks in which the digits 1–9 each appear exactly once. We know that completed Sudoku squares exist, but how many of them will have this additional four-square condition? Adding regions to a Sudoku square creates another search problem of its own; how do we find the solution squares, or even know that any exist?

Taking this question further, how many additional regions of nine cells can we add and still have any Sudoku squares with that property? Four Square Sudoku had four additional regions. Here is one with nine additional regions:

Puzzle 53: Jigsaw Plus.

Complete the grid so that each row, column, and block contains the numbers 1–9 exactly once. In addition, there are nine jigsaw regions on the board in which the numbers 1–9 appear exactly once.

		7					2	
		4	8	3				6
		8						3
1	4	2						
		5				2		
						1	6	5
9						7		
4				9	7	6		
	5					4		

The puzzle above is called Jigsaw Plus rather than just Jigsaw because the jigsaw regions are in *addition* to the usual Sudoku regions, not a replacement for them. In case you are interested, a 9×9 Sudoku square with nine additional nonoverlapping regions is an example of a *multiple Gerechte design*. We're playing these multiple Gerechte designs for fun, but scientists use them to construct balanced, multilayered experiments.

It is possible to have more than nine additional regions, but it is difficult to present such puzzles in a visually pleasing way. So we will stick to nine regions for now. Here is another example, where in this case the nine additional regions are not all contiguous.

Puzzle 54: Rainbow Wrap.

Complete the grid so that each row, column, and block contains the numbers 1–9 exactly once. In addition, there are nine colored bands that wrap around the board. In each diagonal band, the numbers 1–9 appear exactly once. For example, the three yellow cells on the upper left and the six yellow cells on the lower right must together contain the numbers 1–9 exactly once.

							5	9
	9						4	2
		5	9					
		4	2					
					1	5		
					7	8		
1	8						7	
3	5							

Notice that the Rainbow Wrap puzzle has very few clues – only eighteen. In fact, it is worse than you think; the symmetric pair of clues 9 and 7 in the yellow and blue cells along the main, downward diagonal are unnecessary! They can be removed without affecting the puzzle's soundness. We included them only to make the puzzle more accessible and fun to play. But wait – if the minimum number of clues for a sound, symmetrical Sudoku puzzle is conjectured to be eighteen, then how can this puzzle be valid with only sixteen clues? The answer lies in the extra regions. Since the square itself has extra structure, fewer clues are needed to determine the remaining entries. Later in this book we shall see that by adding more and more structure to a Sudoku board, we can actually create puzzles requiring *no* initial clues at all.

All this leads to a different question: Can we add just *any* extra region with nine cells and declare that it is an additional Sudoku region? The answer is no. Some sets of nine cells simply cannot contain 1–9 exactly once, given that we are already guaranteeing that the rows, columns, and blocks of the square are Sudoku regions. Consider, for example, the following hockey stick region:

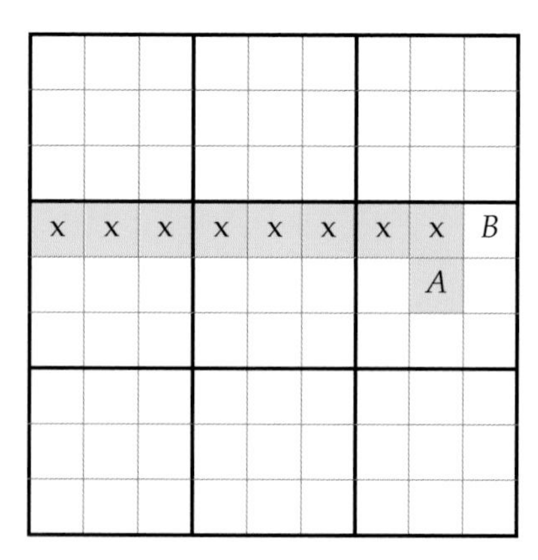

It turns out that there are *no* Sudoku squares in which this particular hockey stick region contains 1–9 exactly once. Here is why: The eight cells marked x must contain eight different digits, each different from the entry at A, since those nine cells make up the hockey stick region. So A is different from all eight xs. Now what can be placed at B? It must be different from all eight of the distinct xs, and also different from A. Sadly, no such digit exists.

When considering the nine-cell subsets permissible as additional regions, do not restrict yourself to connected sets. They can be very disconnected, as in the puzzle below. The extra regions are based on the relative positions of the cells within their 3×3 blocks. These regions do not give as much extra information as the ones in the Rainbow Wrap puzzle. When playing the puzzle, can you tell why?

Puzzle 55: Position Sudoku.

Complete the grid so that each row, column, and block contains the numbers 1–9 exactly once. In addition, in each set of nine same-colored squares, the numbers 1–9 appear exactly once.

							6	8
	3				9			2
		1						7
			4					
				1				
					5			
8						9		
5			3				2	
6	4							

Another way of imposing additional conditions is to have it overlap with other Sudoku boards. Consider, for example, the triple of boards in the next puzzle. Considered separately, the pink grid has 1,977 solutions, the yellow has 2,231, and the blue has 2,181. But only *one* combination adequately accounts for the overlap. From a puzzle-making perspective, it is a small challenge even finding three boards that can overlap in this manner. A far bigger challenge is finding a visually appealing set of initial clues leading to a unique solution. An additional cool feature of this triple puzzle: Each of the three boards has exactly the same rotationally symmetric clue pattern!

Puzzle 56: Venn Sudoku.

Complete the pink, yellow, and blue grids simultaneously so that for every grid, each row, column, and block contains the numbers 1–9 exactly once.

2			1					8						
	3			5										
		7			3	1								
1					5	7			2					1
	5						1			6				
		6	3					9			1	6		
		8	4			9					8	4		
				3			4						2	
5					9			3	1					9
			1					7	8			1		
				2						1			5	
					8	1					9			4
					4	7			5					
							1			9				
			3					6			1			

7

Graphs

Dots, Lines, and Sudoku

To this point, we have been studying Sudoku squares directly, which is to say we have been treating them solely as arrays of symbols. As fruitful as this approach has been, it is hardly the only avenue. An overly literal interpretation can dull the imagination. Sometimes there is insight to be gained from a more abstract approach.

7.1 A PHYSICS LESSON

As a case in point, consider the vexing problem of determining the path of a cannonball. Upon firing the ball, we are confronted with certain important questions. How high will the ball go? Where will it land? We could study these questions at a purely empirical level. That is, we could gather a large number of cannonballs, position our cannon at various angles of inclination, fire the balls, keep track of where they land, and then try to draw some useful generalizations. That seems a bit wasteful, not to mention dangerous.

It was Isaac Newton who solved this problem. His central insight was that this was not a problem about cannonballs at all. Really it was a problem about continuous functions. The ball, you see, does not teleport from one point to another. It moves along a curve, which means the problem of predicting its trajectory is equivalent to the problem of understanding that curve.

So Newton reasoned roughly as follows: After the ball leaves the cannon, the only significant force acting on it is gravity, which we will denote by $-g$. (The use of the negative sign is a notational convention used to indicate that gravity pulls things downward.) Other forces, like wind resistance or the gravitational pull of the moon, are too small to affect the ball's motion in an important way. Now, the key thing about the force of gravity near the surface of the Earth is that it is constant in both magnitude and direction, which is to say it is a vector quantity. Its precise value can be determined experimentally, but for now that is unimportant.

If you have taken a course in basic physics, you might recall that velocity is also a vector quantity. If you are driving in a car your speed might be fifty miles per hour, but your velocity is something like fifty miles per hour due north. A force is something that causes a change in velocity. If there are no forces acting on an object, then its velocity will not change. A change in velocity is referred to as an acceleration.

If the cannonball's acceleration is constant after being fired, then the velocity function has a constant rate of change. The only sort of function with a constant rate of change is a straight line. Straight lines have the form $f(x) = mx + b$, where b represents the point where the curve crosses the y-axis. For our cannonball, the value of b will be the initial velocity of the ball as it emerges from the cannon, which we will denote by v_0.

If we let $a(t)$ and $v(t)$ represent the acceleration and velocity of the ball at time t, then we now have two equations with which to work:

$$a(t) = -g \qquad \text{and} \qquad v(t) = -gt + v_0.$$

Velocity measures the rate of change in position. We knew that the velocity function was a straight line because its rate of change was constant. But what kind of function has a linear rate of change? It is a result of elementary calculus that the parabola is the only such function. If we let $x(t)$ denote the position of the ball at time t, and if we let x_0 denote the initial position of the ball, then we have

$$x(t) = -\tfrac{1}{2}gt^2 + v_0t + x_0.$$

Just like that we have deduced that the cannonball traces out a parabola as it moves. We did not need to fire a single ball to arrive at this conclusion. We needed some basic calculus and the observation that the ball's acceleration was constant.

With similar reasoning, you can answer almost any question about the trajectory of the cannonball. For example, the highest point reached by the ball corresponds to the point on the parabola at which the tangent line is horizontal. In the language of calculus, we seek the point at which the derivative is zero.

Physical questions about the path of the cannonball are readily translated into mathematical questions about the properties of parabolas. And since continuous functions are easier to study than physical objects, we think you will assent to the usefulness of this approach.

7.2 TWO MATHEMATICAL EXAMPLES

For some purely mathematical examples of the same general phenomenon, let us consider the interrelationship between geometry and algebra.

Remember FOIL? That was the acronym you learned for multiplying binomials. It stands for, "First, Outer, Inner, Last," which makes sense when you recall that

$$(a + b)(c + d) = ac + ad + bc + bd.$$

In your algebra class, you learned this as a rule for manipulating symbols. It was a consequence of the distributive and commutative properties of addition and multiplication. Very nice. But the product of $(a + b)$ with $(c + d)$ also represents the area of a rectangle with length $(a + b)$ and width $(c + d)$. We can use this to obtain a visual presentation of our algebraic fact:

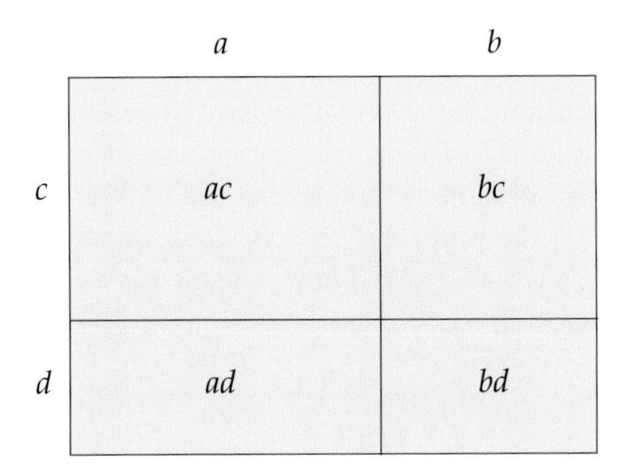

Somehow seeing the picture really makes things come alive; the area of the large rectangle with side lengths $a + b$ and $c + d$ is the sum of the areas of the four small rectangles with areas *ac*, *ad*, *bc*, and *bd*. Pictures like this give us hope that we are not just manipulating symbols according to arbitrary rules. We are saying something intelligible about real-world objects.

Many algebraic identities can be given new life through a geometric presentation. For example, the rule that

$$x^2 - y^2 = (x + y)(x - y)$$

makes perfect sense when you ponder the following diagrams:

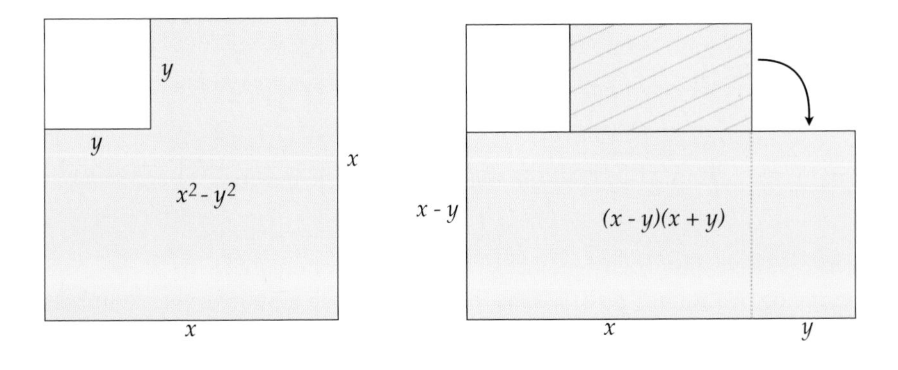

You might enjoy devising your own diagrams for other algebraic identities. Do not limit yourself to two dimensions. Algebraically speaking we know that

$$(x+y)^3 = x^3 + 3x^2y + 3xy^2 + y^3.$$

But $(x+y)^3$ is also the volume of a cube with side length $(x+y)$. If you slice up such a cube in a clever way, then you can see why the algebraic identity is correct.

Recasting algebraic facts in the language of geometry was a good idea. So is using techniques from other branches of mathematics to solve geometrical problems.

For example, how do we find the area of a circle? You are surely aware that for a circle with radius r we have $A = \pi r^2$, but from where does that formula come? Archimedes devised a purely geometrical derivation around 225 BC. His method was ingenious, but becomes rather long when written out in full detail. (We recommend the book by Dunham [21] if you are interested.)

While Archimedes' solution can still be enjoyed both for its cleverness and its historical significance, the problem becomes straightforward if you are familiar with the techniques of calculus. We know that the top half of a circle of radius r centered at the origin of a coordinate axes is given by the following equation and graph:

$$y = \sqrt{r^2 - x^2}$$

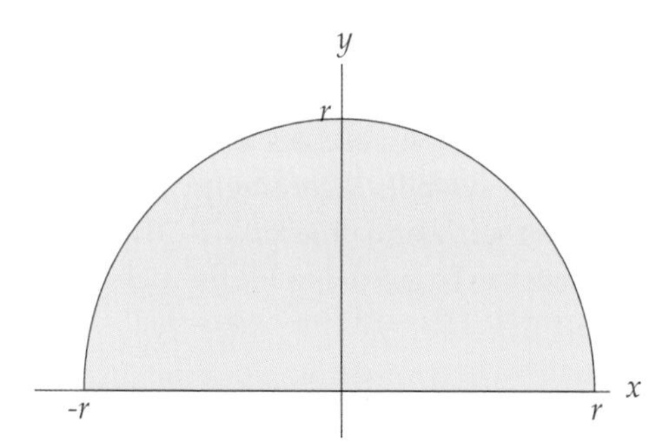

The method of *integration* from calculus allows us to find the area between such a curve and the horizontal axis. The area of the full circle will be twice the area under the curve. The so-called *integral* for this area can be evaluated by an algebraic strategy known as trigonometric substitution, or by looking it up in a table of integrals. If we let $A(r)$ denote the area of the circle, then we obtain

$$A(r) = 2\int_{-r}^{r} \sqrt{r^2 - x^2}\,dx = 2\left[\frac{x}{2}\sqrt{r^2 - x^2} + \frac{r^2}{2}\sin^{-1}\frac{x}{r}\right]_{-r}^{r} = \pi r^2.$$

As we have just seen, geometry and algebra often meet at π. So before continuing, we cannot resist a short break to include a Sudoku puzzle in honor of that great number:

Puzzle 57: Jigsaw Pi Sudoku.

Each row, column, and jigsaw region must contain exactly the first twelve digits of π, including repeats: 3.14159265358. Notice that each region will contain two 1s, two 3s, three 5s, and no 7s.

3			1	5	4			1		9	5
	1			3					1	3	6
		4			3		8			2	
5			1			9	2	5			1
	9			5			5				
5	8	1			9			3		6	
	5		8			2			5	5	3
				5			6			1	
2			5	1	5			5			9
	6			4		1			3		
1	5	1					5			5	
5	5		4			3	1	6			8

It is a common strategy in mathematics to answer a question from one field by translating it into a question in another. The first step in solving a math problem often involves recasting it in a different form. Mathematicians carry as part of their standard toolkit a large collection of abstract structures and problem-solving techniques. Confronted with a new problem, we check to see if anything in our toolkit is potentially useful. Let us see if we can apply this strategy to Sudoku.

7.3 SUDOKU AS A PROBLEM IN GRAPH COLORING

It was not long after Sudoku arrived on the scene that mathematicians noticed the puzzle could be recast as a problem in graph coloring. We saw an example of graphs back in Chapter 1, when we considered the Bridges of Königsburg. There we saw that a graph consisted of a set of vertices, some of which were connected to others by edges. We should note that there is no requirement that the edges be depicted as *straight* lines. We only care about what is connected to what, and a curve can make that relationship just as clear as a line.

This is certainly a different notion of a graph from what students learn in precalculus and calculus. In those courses, it is routine to speak of "the graph of a function," but that is something else entirely from what we are considering here. The English word *graph* comes from the Greek word *graphein,* which means "to write." In general, we apply the term to any sort of visual display of an abstract object. We speak of "computer graphics," for example, in referring to visual presentations of image data. The graph of a function permits us to visualize its behavior in ways that are very difficult from the formula alone.

Diagrams similar to what we are calling graphs have been used for centuries to depict the physical arrangements of atoms in molecules. Such diagrams used to be referred to as *chemicographs.* For example, here is the chemicograph for the caffeine molecule:

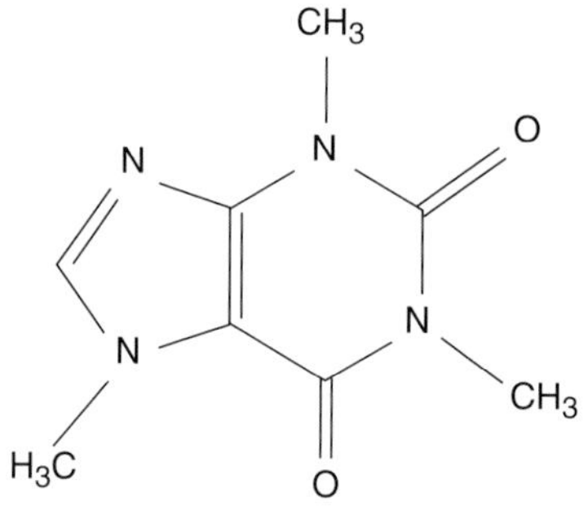

Writing in 1878, the American mathematician J. J. Sylvester [42] noticed a possible connection between such diagrams and certain problems in algebra on which he was working. He shortened the term "chemicograph," to "graph," and the name stuck. There is some irony in this, since his proposed connection between chemistry and algebra turned out to be a dead end [10].

How might we associate a graph to a Sudoku square? As usual, let us start with the smaller and more manageable case of 4×4 Shidoku. Think of each of the sixteen cells in the grid as a vertex. It will be convenient to make no distinction between a vertex in the graph and the cell it represents in the Shidoku puzzle. For example, we will speak casually of the digit in a vertex even though it would be more correct to say, "the digit contained in the cell represented by the vertex."

Each of these vertices generates a zone consisting of the three other vertices in its row, the three other vertices in its column, and the one other vertex in its block that is not in the same row or column, as shown below left. For each vertex, add connecting edges to every other vertex in its zone. For example, below right we see the part of the Shidoku graph that is edge-connected to the upper-left cell of the Shidoku square. Two vertices share an edge exactly when their corresponding cells share a row, a column, or a block on the Shidoku square.

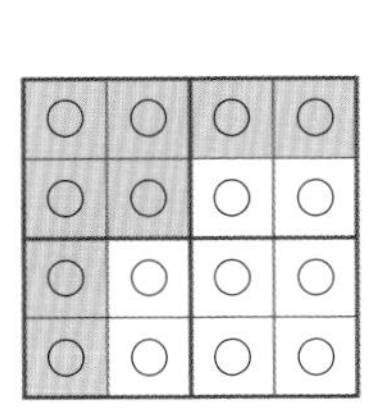 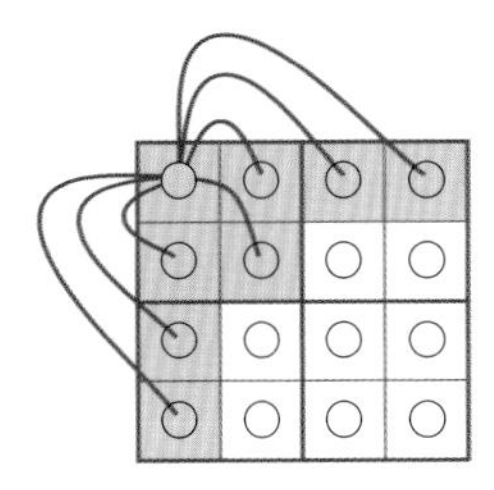

The finished Shidoku graph has sixteen vertices. Each of these vertices has seven edges coming out of it. Even for this small 4×4 Shidoku case the graph is messy; imagine a seven-legged "spider" such as the one shown above right, sitting on each cell of the board. The graph has sixteen of these seven-legged spiders, where each "leg" or edge is be shared by two spiders. The total number of edges in the Shidoku graph is therefore $\frac{(16)(7)}{2} = 56$.

For 9×9 Sudoku, the situation is far worse. Each of the 81 vertices is connected by an edge to 8 vertices in its row, 8 vertices in its column, and 4 other vertices in its block not in the same row or columns. Just imagine those 20-legged spiders! The graph has 81 of them, with each leg shared by two spiders, giving a total of $\frac{(81)(20)}{2} = 810$ edges for the Sudoku graph. For the moment, we will confine our discussion to the simpler 4×4 Shidoku case.

Cells appearing in the same Shidoku zone must contain different digits. We shall represent this fact in the graph by assigning colors to the vertices, with connected vertices receiving different colors. Let us introduce some terminology. Two vertices on a graph are said to be *adjacent* if they are connected by an edge. A graph is said to be *properly colored* if adjacent vertices are always assigned different colors. If a proper coloring uses n colors, then we shall refer to it as a *proper n-coloring*.

Given a graph, we might now ask how many colors are needed for a proper coloring. Some graphs can be properly colored with three colors, for example, while others cannot. Two of the three graphs in the puzzle below are properly three-colorable, and one is not. If you are interested, the first graph is known as the *Petersen graph*, the second is the graph determined by the edges of a dodecahedron, and the third is known as the *Grötzsch graph*.

Puzzle 58: Three-Coloring Graphs.

Color the vertices of each graph so that adjacent vertices receive different colors. Try to do this with only three colors. You will find that one of the graphs can never be properly colored with three colors; but which one?

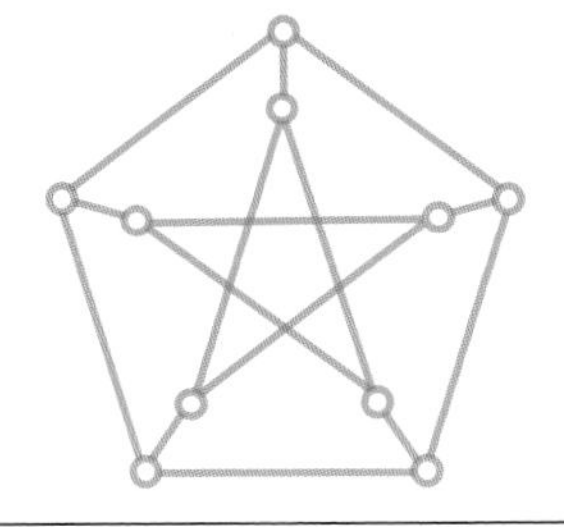 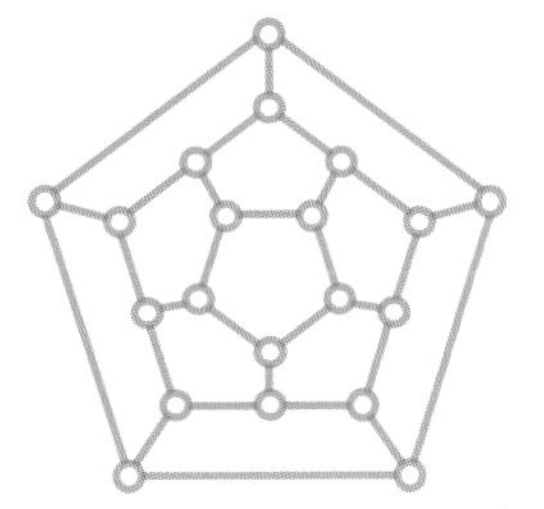 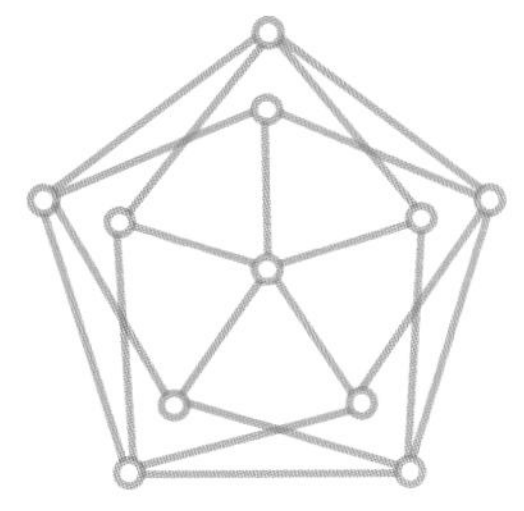

There is a perfect correspondence between a proper four-coloring of a Shidoku graph and a completed Shidoku square. The different colors assigned to adjacent vertices in the graph correspond to the different digits placed in the cells of the same region. For example, consider our spider from the first cell of the Shidoku board, and choose the proper coloring shown below left. If 1=red, 2=green, 3=purple and 4=blue then this coloring corresponds to filling in the first row, block, and column of the Shidoku square as shown below right.

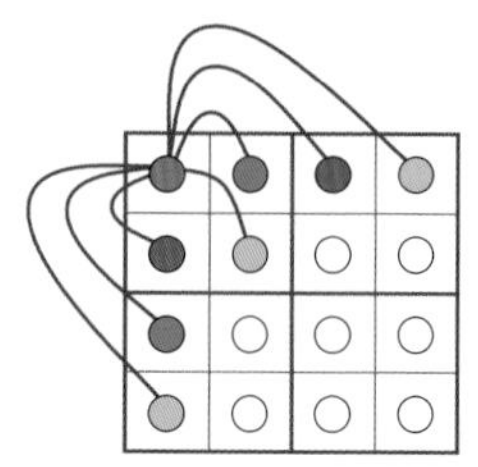

1	2	3	4
3	4		
2			
4			

Similarly, a Shidoku puzzle can be thought of as a partial coloring of the Shidoku graph, where the initial clues in the puzzle correspond to vertices on the graph that have already been colored. In the puzzle, the solver must determine how to place digits in the remaining cells. In the graph, the solver must determine how to color the remaining vertices. If the puzzle is sound, then there will be only one way of doing so.

The minimal number of colors needed for a proper coloring is known as the *chromatic number* of the graph. The Petersen graph in Puzzle 58, for example, has chromatic number 3, since its vertices can be properly colored with three, but not two, colors. On the other hand, the Grötzsch graph from that same puzzle has chromatic number 4, since it can be properly colored with four, but not three, colors.

The chromatic number of the Shidoku graph is 4. To see this, note that any valid Shidoku square represents a proper coloring of the graph using four colors. This tells us that the chromatic number is no more than 4. But the four vertices in any block of the square are all connected to each other, meaning that we need at least four colors just for those four vertices.

By a similar argument, the chromatic number of the Sudoku graph is 9.

Thus, our transformation is complete. What started as a question about an array of numbers is now a question about coloring the vertices of a graph. Anything we can say about the Shidoku or Sudoku graphs translates into a result about Shidoku or Sudoku squares and vice versa.

Let us conclude this section with a few more graph-coloring brainteasers. Though our interest in this section involves vertex-coloring problems, it can also be amusing to contemplate edge-colorings.

Puzzle 59: Edge-Coloring Graphs.

Consider again the three graphs below. This time, color the edges of each graph so that edges meeting at a vertex are assigned different colors. What is the smallest number of colors needed for a proper edge-coloring in each case? (Hint: One graph requires three colors, one requires four, and one requires five.)

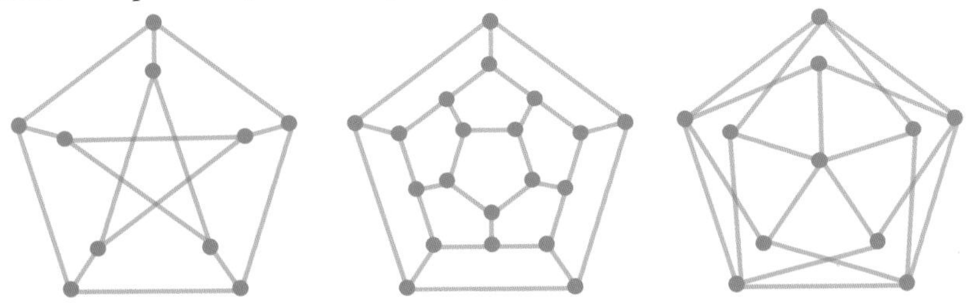

For the last brainteaser we need another definition. A graph in which every vertex is connected to every other vertex is called a *complete graph*. We use the notation K_n to denote the complete graph on n vertices. What sorts of things happen when we color the edges of complete graphs with only two colors? For n higher than 3, such two-colorings will not be proper. This is clear, since every vertex in the complete graph K_n connects to $n - 1$ other vertices. Thus, a proper edge-coloring of K_n requires at least $n - 1$ colors.

Puzzle 60: Triangles in Complete Graphs.

The graph on the left is K_5, the complete graph on five vertices. The graph on the right is K_6, the complete graph on six vertices. First, color each edge of K_5 either red or blue in such a way that there are no all-red triangles and no all-blue triangles. Then argue that no matter how the edges of K_6 are colored with red or blue, there must *always* be an all-red triangle or an all-blue triangle.

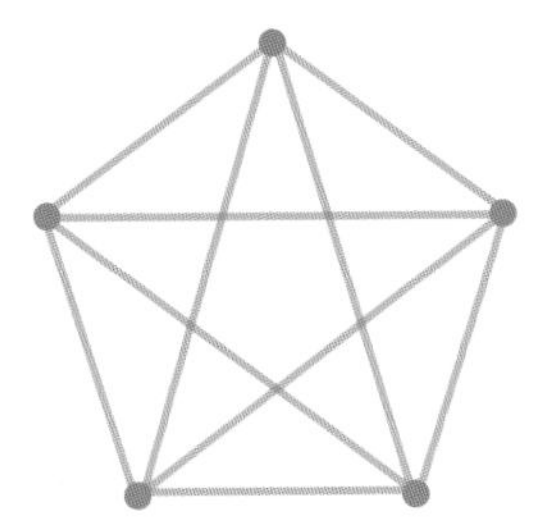

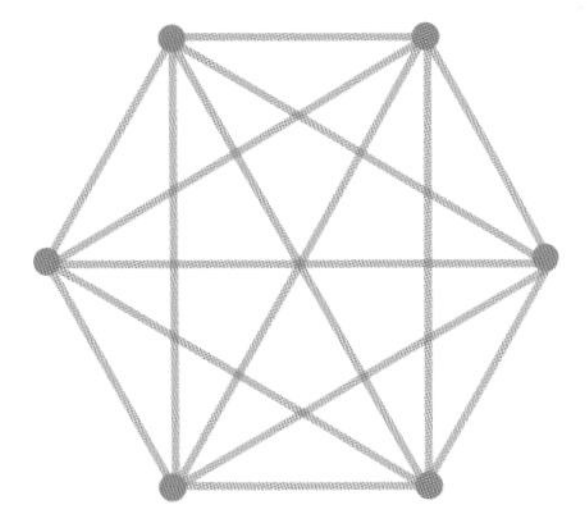

7.4 THE FOUR-COLOR THEOREM

Every culture has its unwritten rules. Little bits of wisdom everyone knows but no one talks about. Mathematics is no different, and we must now include this short section to avoid violating a big one. The rule says that any discussion of graph coloring must include a reference to the Four-Color Theorem.

In 1852 a fellow named Francis Guthrie attempted to color the counties in a map of England. To make them easily distinguishable, he desired that neighboring counties be given different colors. He discovered that four colors were sufficient for this purpose. But what of other maps? Here's one to try:

Puzzle 61: Four-Coloring America.

Color this current map of the mainland United States with four colors so that any two states that share a boundary are different colors. Can you do it with just three colors?

So the counties of England can be four-colored, as they were in 1852 at any rate, and the states of America can be four-colored. Guthrie was unaware of any map that required more than four colors, but maybe he just lacked imagination. Could any such map exist?

Guthrie's musings might have been the end of it, but he happened to mention the problem to his brother, Frederick. At that time, Frederick was a student of the British mathematician Augustus De Morgan. Unable to resolve the question himself, De Morgan mentioned it in correspondence with another British mathematician, William Hamilton. Eventually the problem found its way to Arthur Cayley, who gave the problem its first publication in a scholarly venue in 1879.

At first it seemed the problem would fall quickly. Shortly after the publication of Cayley's paper, Alfred Kempe devised a widely acclaimed proof. The following year, a different proof was offered by Peter Tait. Alas, each had made subtle, but fatal, errors. Kempe's proof fell in 1890 and Tait's succumbed in 1891. Nobody could find a map that needed more than four colors, but neither could anyone prove that such a map was impossible. Four colors certainly seems reasonable, at least in light of the following simple examples:

Puzzle 62: Four-Color Wheels.

Consider the special case where counties on a map are in the "wheel" configurations shown below, with one county in the center surrounded by some number of other counties. What is the smallest number of colors needed to color the regions of each wheel-map so that regions sharing a border are colored differently? Do you see a pattern?

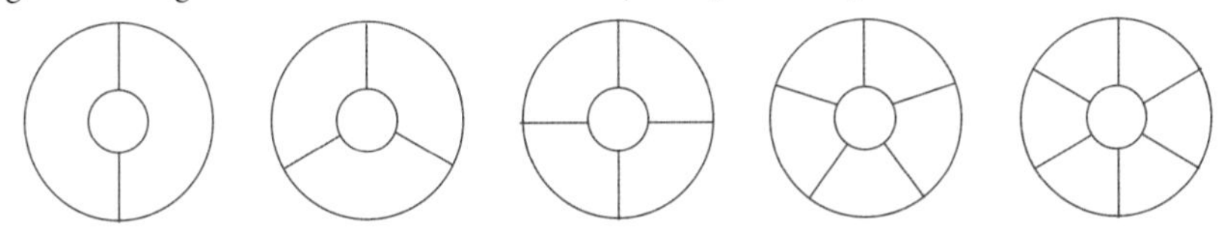

As we discussed previously, the problem was finally conquered in 1976, when Kenneth Appel and Wolfgang Haken produced a proof. Or did they? Their argument relied on the efforts of a computer, you see. It was very similar to our work in Chapter 4 in counting the number of distinct Sudoku boards. Great ingenuity was required to reduce the problem to something computationally tractable, but the fact remains the computer did the heavy lifting. This was the first example of a significant, computer-assisted proof. As we mentioned, most mathematicians are happy to accept the computer proof as a demonstration of the theorem's correctness. The problem is that the computer verifies without clarifying. A noncomputer proof of the result would be a marvelous thing.

If you are interested in pursuing this subject further then we recommend the excellent book by Robin Wilson, *Four Colors Suffice* [46]. For our part, we will simply hope the math gods regard this discussion as sufficient propitiation and will now return to our discussion of Sudoku graphs.

7.5 MANY ROADS TO ROME

Now for the bad news. Interesting though it is to express a Sudoku puzzle in the form of a graph-coloring problem, it is unclear if any additional insight is gained by doing so. We are unaware of any result about Sudoku puzzles provable using the techniques of graph-coloring that could not have been arrived at more easily by other means.

Does that mean we wasted our time? Certainly not! For one thing, short of making an outright error in your reasoning, it is rare that you chase down a *completely* dead end. You always end up learning *something*. Noting a connection between Sudoku puzzles and a well-established branch of mathematics is, all by itself, a useful observation.

But there is more. While it is true that the graph-coloring approach has not led to novel insights into Sudoku puzzles, it is also true that we can gain new perspective on familiar results. We have a specific example in mind, but it requires some set-up.

Among the biggest open questions about Sudoku is the minimal number of starting clues required to ensure a unique solution. We think of this as the "rock star" problem in Sudoku studies, because many people want to be the one to earn the fame and glory of solving it. It is easy to state but hard to solve. We have already mentioned there are many examples of sound seventeen-clue puzzles, but no known examples of sound sixteen-clue puzzles. This makes people suspect that seventeen is the minimal number, but a suspicion is not a proof. We will have more to say about this in Chapter 9.

If it is too difficult to determine the minimal number exactly, perhaps we can at least establish a lower bound. In the previous chapter, we mentioned that a sound Sudoku puzzle must have at least eight starting clues, but now we look more closely at that fact. The following puzzle nicely illustrates why seven starting clues are insufficient for a sound puzzle.

Puzzle 63: Seven Isn't Enough.

Complete the grid so that each row, column, and block contains the numbers 1–9 exactly once. You should find that there are *two* possible ways of doing this. Notice that there is no way that this "puzzle" could have a unique solution, since no 8's or 9's are given as clue values. The two possible solutions to this puzzle are in the text below.

		3	7					
	2	7		3			5	
					5		3	4
		4			2			1
	5			1			7	
6			4			3		
7	4		2					
	1			5		6	4	
					6	1		

Since this psuedo-puzzle uses only seven of the nine possible clue values, it cannot have a unique solution. Any solution of the puzzle could be modified by swapping 8s and 9s to produce another valid solution. In this case, there happen to be exactly two solutions, differing only in the positions of the 8s and 9s:

5	9	3	7	6	4	2	1	8
4	2	7	1	3	8	9	5	6
1	6	8	9	2	5	7	3	4
9	3	4	5	7	2	8	6	1
8	5	2	6	1	3	4	7	9
6	7	1	4	8	9	3	2	5
7	4	6	2	9	1	5	8	3
3	1	9	8	5	7	6	4	2
2	8	5	3	4	6	1	9	7

5	8	3	7	6	4	2	1	9
4	2	7	1	3	9	8	5	6
1	6	9	8	2	5	7	3	4
8	3	4	5	7	2	9	6	1
9	5	2	6	1	3	4	7	8
6	7	1	4	9	8	3	2	5
7	4	6	2	8	1	5	9	3
3	1	8	9	5	7	6	4	2
2	9	5	3	4	6	1	8	7

So it is easy to show that a valid Sudoku square must have at least eight starting clues. This result generalizes easily to a Sudoku square of size $n \times n$, and it tells us that the minimal number of starting clues is at least $n - 1$.

An alternative proof of this result was provided by Herzberg and Murty [24]. Compared to the previous proof, this one will use some pretty serious machinery. We will need a few facts about polynomials to understand it. As before, if you find yourself getting bogged down in the details you can skip ahead without losing anything.

We have previously mentioned the idea of factoring a quadratic equation to find its roots, that is, the values of x that make it equal to 0. This works because every root of a polynomial corresponds to a factor, and every factor corresponds to a root. For example, if we have the equation

$$3x^2 - 13x - 10 = 0,$$

then we can factor it as $(x - 5)(3x + 2)$ to conclude that the solutions are 5 and $-\frac{2}{3}$.

For a more complex example, consider this:

$$2x^3 - 13x^2 - 10x + 21 = 0.$$

With a little trial and error, you might discover that $x = 1$ is a solution to this equation. That tells you that $x - 1$ must be a factor of the polynomial. That is,

$$2x^3 - 13x^2 - 10x + 21 = (x - 1)q(x),$$

where $q(x)$ is some yet-to-be-determined quadratic polynomial. We could employ polynomial long division to obtain

$$\begin{aligned} 2x^3 - 13x^2 - 10x + 21 &= (x - 1)(2x^2 - 11x - 21) \\ &= (x - 1)(x - 7)(2x + 3), \end{aligned}$$

from which we conclude that the solutions are $x = 1, 7, -\frac{3}{2}$.

Armed with those facts, let us return to our alternate proof. Hurzberg and Murty [24] began by defining a function as follows: Let G be a graph and let C be a partial proper coloring of the graph. Let x be a positive integer. Then we define $p_{G,C}(x)$ to be the number of ways of using no more than x colors to complete the coloring of the graph. Though it is not easy to prove, it can be shown that this function is actually a polynomial (as opposed to some other, more exotic, kind of function). Even better, it turns out that this polynomial has integer coefficients and a leading coefficient of 1.

Now let G be the Sudoku graph and let C be given by the set of starting clues. Since the chromatic number of G is 9, we know that G cannot be colored with fewer than 9 colors. That means that $p_{G,C}(x) = 0$ if $1 \leq x \leq 8$. In other words, the number of ways of completing the coloring using eight or fewer colors is 0. As we have seen, every root of the polynomial corresponds to a factor and vice versa.

Let us denote by s the number of digits having representatives among the starting clues. Then we must have

$$p_{G,C}(x) = (x - s)[x - (s + 1)][x - (s + 2)] \dots (x - 8)q(x),$$

where $q(x)$ is a polynomial with integer coefficients. Since the polynomial takes on the value 0 when x is smaller than 8, we must have one factor for each of these numbers.

That last step was a bit abstract, so let us consider a more concrete example. Suppose that only five of the nine digits are represented among the starting clues of a Sudoku puzzle. Then we have $s = 5$. We know that the graph cannot be colored with either five, six, seven, or eight colors. That means $p_{G,C}(x)$ is equal to zero when $x = 5, 6, 7$, or 8. Therefore $(x - 5)$, $(x - 6)$, $(x - 7)$, and $(x - 8)$ are all factors of the polynomial. The result of dividing $p_{G,C}(x)$ by those four factors is some other polynomial, which we shall call $q(x)$. That gives us

$$p_{G,C}(x) = (x - 5)(x - 6)(x - 7)(x - 8)q(x).$$

We do not know precisely what polynomial $q(x)$ is, but we do know it has integer coefficients. That is all we need.

Now for the punchline. A sound Sudoku puzzle is one for which $p_{G,C}(9) = 1$. Expressed in words, this formula says there is only one way of completing the partial coloring using nine colors. But we also know that

$$p_{G,C}(9) = (9 - s)[9 - (s + 1)][9 - (s + 2)] \dots (9 - 8)q(9) = (9 - s)!\, q(9).$$

If $s \leq 7$ then the right-hand side of this equation is greater than 1. That means it is impossible to have a unique solution to your Sudoku puzzle unless there are at least eight digits represented among the starting clues. Precisely as we saw before.

Why have we belabored this? This proof is terribly clever, but does it not seem like an awful lot of work to prove something we already knew? Perhaps it was, but permit us a few words in its defense. First, we should mention that the idea of studying the coloring properties of graphs by attaching polynomials to them is part of the standard toolkit for researchers in graph theory. The function $p_{G,C}$ is a small variation on a standard device known as the *chromatic polynomial* of a graph. For people already familiar with such things this proof seems fairly natural.

But there is a more important point to be made, and it relates to something we discussed earlier. Good proofs do not just verify, they also clarify. A proof's conclusion is hardly the only important thing about it. Often it is not even the most interesting part. Multiple proofs of the same result can be useful because each proof shows you relationships among mathematical objects you might not previously have noticed.

It is rather like moving to a new town. Initially you do not know your way around very well, but you have certain business you must transact. You need to find routes to work, to the market, to the bank, that sort of thing. So you learn a few of the major roads and are then able to do what you need to do.

That is the beginning, not the end, of your pursuit of wisdom. What you really want is a mental map of all the roads. As you spend more time in the town, you gradually learn more and more of the terrain. Maybe one day there is major road construction on Main Street, so you consult your mental map to find an alternate route. Maybe the shortest route normally involves plowing down Main Street, but you know there is heavy traffic there at midday. So you use the side streets. A thoroughgoing knowledge of the terrain provides you with multiple routes to the same destination. That can be very useful.

So it is with mathematics. The abstract objects we study have connections and interrelations that can be difficult to fathom. When we prove a result, we have illuminated a small part of the terrain. An alternate proof of the same result illuminates still more. The results correspond to our various destinations. But our real interest is in the terrain.

7.6 BOOK EMBEDDINGS

Our Sudoku graphs give us an alternate way of thinking about Sudoku squares and puzzles, but the graphs are very messy: many vertices, many edges, all interacting in ways that are difficult to visualize. It is necessary to impose some organization on this complexity. Toward that end let us ponder the full Shidoku graph. On the left, the graph is shown with our original seven-legged spider in red. On the right, we see the same graph with some helpful coloring added. In both graphs, we label the vertices from 1 to 16 for easy reference.

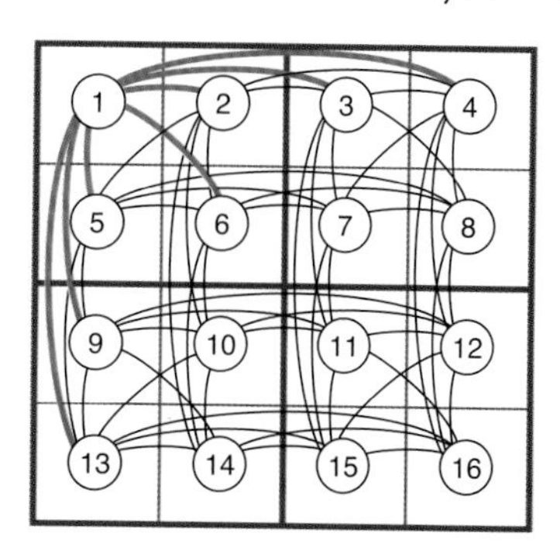

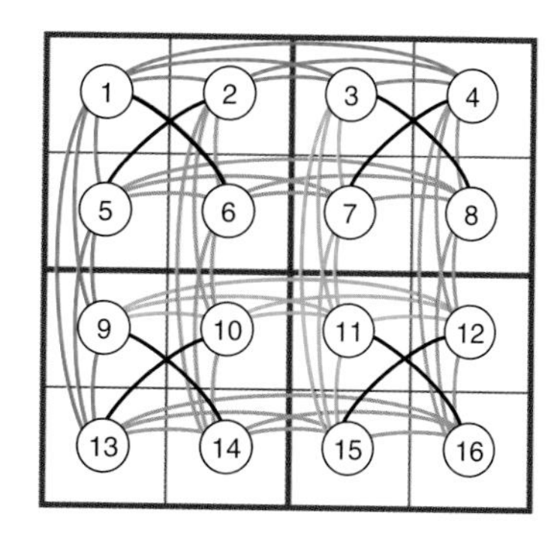

Dealing with such complexity is a standard problem in graph theory. We define graphs to provide visual presentations of complex interrelationships. If the graph is just a riot of crooked lines zigzagging across the page, then it is hard to see why we bothered.

Consequently, graph theorists devote a fair amount of effort to questions regarding the visual presentation of graphs. We ask questions like: Is it possible to draw the graph in a plane so that none of the edges cross? If it is not possible, then what is the smallest number of crossings that is necessary?

One gadget that has been devised along these lines is the notion of a *book embedding*. You begin by lining up the vertices in an order of your choice. This forms the spine of the book. The pages then contain the individual edges, drawn so that on any given page the edges do not cross. The number of pages needed for this depends on the manner in which we have ordered the vertices and also on the manner in which we draw the edges. The smallest number of pages needed, among all the possible orderings of the vertices and all the possible drawings of the edges, is referred to as the *book thickness* of the graph. It is denoted $BT(G)$ (read, "Bee Tee of G.")

A simple example should clarify things. Consider a graph G with four vertices, numbered 1–4, with vertex 1 connected to vertex 3 and vertex 2 connected to vertices 3 and 4. Those are the only edges. One way of drawing the graph G is shown below.

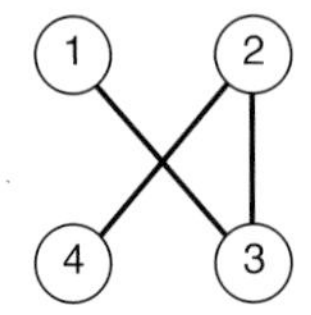

We could embed this graph in a book with two pages as shown below; imagine that the spine of the book is along the dotted line, with one page above the line and one page below. Ordered in this way, we can draw the edges connecting 1 to 2 and 2 to 3 without crossing. They can go on the same "page." But it is impossible to connect vertex two to vertex four without crossing a previous edge. Thus, the edge connecting two to four must go in a different page.

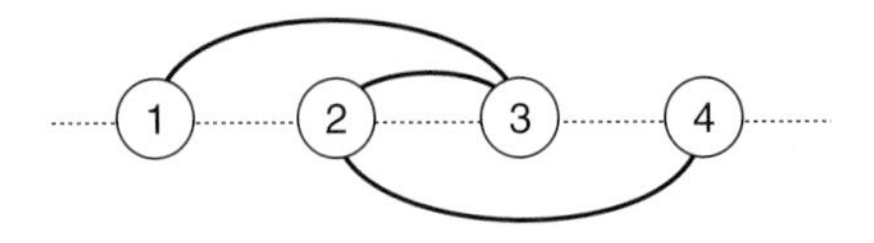

On the other hand, a different ordering of the vertices along the spine permits a one-page embedding, showing that the book thickness of our graph G is just $BT(G) = 1$:

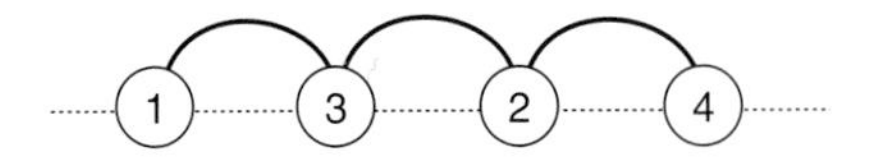

Some attention has been paid to book embeddings of Sudoku graphs. For convenience, we shall focus on the more manageable Shidoku (4×4) graphs. In work by Blankenship and Music [11], it was shown that if S_4 denotes the Shidoku graph, then we have $4 \leq BT(S_4) \leq 6$. The lower bound of 4 can be established via elementary results in graph theory. The upper bound of 6 is established by exhibiting a six-page embedding of the graph.

Let us see how that is done. Fixing a certain order along the spine, the first and second pages record the edge data from all four rows of the Shidoku square. In the diagram below, the colors are taken from the full graph shown earlier. The first page consists of all the edges above the line, while the second page contains all edges below the line.

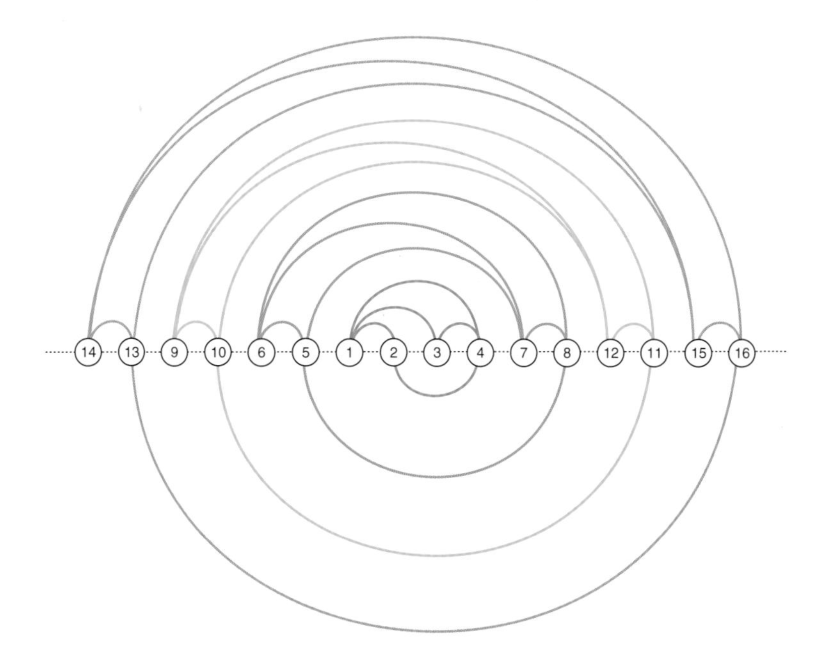

Keeping the same vertex order along the spine, pages 3 and 4 describe the first and third column relationships of the Shidoku square:

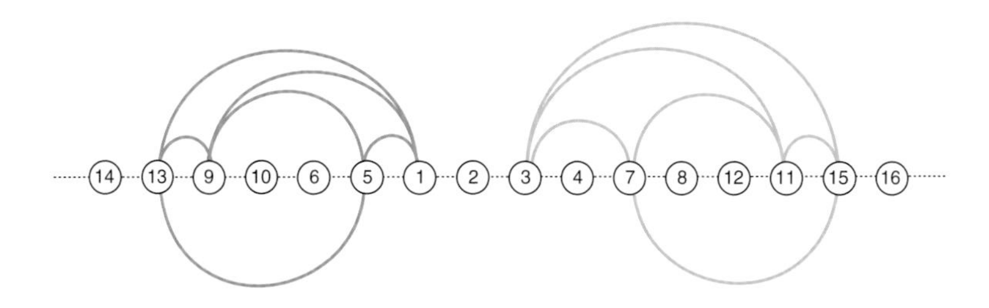

Blankenship and Music [11] then managed to fit both the second and fourth column information *and* the remaining block relationships (in black) on the fifth and sixth pages:

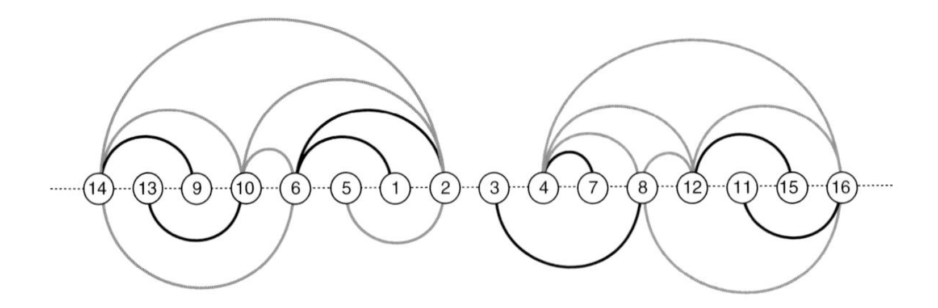

Elegant! Is there an even more clever ordering of vertices that allows a five-page embedding? If you manage to answer that question let us know.

If this all seems a bit esoteric and rarefied to you, then rest assured it does to us as well. But is that a criticism? Not everything in life is about building a better mousetrap. Mathematicians solve problems for the same reason people go exploring in forests and caves. To see what is out there.

To the person who asks, "Who cares about book embeddings of Sudoku graphs?" two replies are possible. The highly practical side of mathematics would stress that one never knows from where the next great idea is coming. The history of our discipline is littered with examples of objects first studied on the whim of some curious mathematician, which later turned out to be critically important to more practical concerns.

That is certainly true and important, but for the moment we would stress a different point. Sudoku puzzles are interesting. Period. Our graphs arose naturally upon thinking about Sudoku, and they are interesting for that reason. Book embeddings help us understand these graphs. That which helps us understand an interesting object is, itself, interesting.

8

Polynomials

We Finally Found a Use for Algebra

Graphs are not the only mathematical objects that we can associate with Sudoku squares. There are also polynomials to consider. Just as we were able to use the language of graph coloring to capture the relationships among the cells in a Sudoku square, so too can we use polynomials for the same purpose. Once again we shall focus on the case of Shidoku to keep things to a manageable size. The arguments we shall present are developed in greater detail in the paper by Arnold, Lucas, and Taalman [6].

8.1 SUMS AND PRODUCTS

If you have ever used algebra to solve a word problem then you are familiar with the idea that equations can be used to capture relationships among numbers. Suppose we tell you that John, Kate and Mary have ages that sum to 73. Suppose further that Kate is twice as old as Mary but three years younger than John. Asked to find their ages you would probably declare that if x denotes Mary's age, then Kate's age is $2x$ and John's age is $2x + 3$. Then we have

$$x + 2x + (2x + 3) = 5x + 3 = 73.$$

This equation captures all of the relevant information from the problem. It is easily solved to show that Mary is 14, Kate is 28, and John is 31.

Each of the cells in a Shidoku square contains a particular digit from 1 to 4. Cells in the same Shidoku region must be distinct from one another. It ought to be possible to devise equations expressing these facts in mathematical form. If we are successful, then our equations will contain all of the important information from the square itself.

To see how this could work, consider the most basic fact of them all. Each cell contains either a 1, 2, 3, or 4. If w represents one of the sixteen cells on a Shidoku board, how do we write an equation that says, "The cell represented by w is equal to exactly one of those four digits?" If you recall what we said in the last chapter regarding roots and factors in polynomials, then you might come up with this:

$$(w-1)(w-2)(w-3)(w-4)=0.$$

Each of the sixteen individual cells must satisfy this polynomial equation. If we imagine using a different letter for each cell, then we have sixteen different equations.

What about the relationships among the variables in the same Shidoku region (a row, column, or 2×2 block)? If we let w, x, y, and z denote the four cells in some region, then we know that all four of them must take on different values. This means one of them is 1, one of them is 2, one of them is 3, and one of them is 4, so their sum is $1+2+3+4=10$. In other words,

$$w+x+y+z=10.$$

This is necessary but not sufficient. The four cells in any given region do add up to ten, but this equation by itself does not ensure that the four numbers are different. As far as this equation is concerned, we could have $w=x=1$ and $y=z=4$.

What else can we say? Well, the product of 1, 2, 3, and 4 is 24, which means that

$$wxyz=24.$$

It turns out that if four numbers taken from 1, 2, 3, and 4 satisfy both of the equations above, then those numbers must all be different. The only way to choose four numbers from among 1, 2, 3, and 4 that sum to 10 and multiply to 24 is to choose each number exactly once. With these two equations we have perfectly encoded one rule of Shidoku: each region has no repeated numbers.

This gives us two additional polynomial equations for each of the Shidoku regions. Since there are four rows, four columns, and four 2×2 blocks, we have a total of twelve regions and twenty-four more equations. Together with the sixteen we had before that makes forty equations in sixteen variables. Every solution to those forty equations represents a Shidoku square, and every Shidoku square represents a solution to those forty equations.

Let's make sure we know what we are talking about. Suppose the cells on a Shidoku board are labeled like this:

a	b	c	d
e	f	g	h
i	j	k	ℓ
m	n	o	p

Then assigning values to these sixteen letters to make a Shidoku square is precisely equivalent to finding values for $a, b, c, \ldots, p$ that satisfy the following forty equations:

$$\begin{array}{rrr}
(a-1)(a-2)(a-3)(a-4)=0 & a+b+c+d=10 & abcd=24\\
(b-1)(b-2)(b-3)(b-4)=0 & e+f+g+h=10 & efgh=24\\
(c-1)(c-2)(c-3)(c-4)=0 & i+j+k+\ell=10 & ijk\ell=24\\
(d-1)(d-2)(d-3)(d-4)=0 & m+n+o+p=10 & mnop=24\\
(e-1)(e-2)(e-3)(e-4)=0 & & \\
(f-1)(f-2)(f-3)(f-4)=0 & a+e+i+m=10 & aeim=24\\
(g-1)(g-2)(g-3)(g-4)=0 & b+f+j+n=10 & bfjn=24\\
(h-1)(h-2)(h-3)(h-4)=0 & c+g+k+o=10 & cgko=24\\
(i-1)(i-2)(i-3)(i-4)=0 & d+h+\ell+p=10 & dh\ell p=24\\
(j-1)(j-2)(j-3)(j-4)=0 & & \\
(k-1)(k-2)(k-3)(k-4)=0 & a+b+e+f=10 & abef=24\\
(\ell-1)(\ell-2)(\ell-3)(\ell-4)=0 & c+d+g+h=10 & cdgh=24\\
(m-1)(m-2)(m-3)(m-4)=0 & i+j+m+n=10 & ijmn=24\\
(n-1)(n-2)(n-3)(n-4)=0 & k+\ell+o+p=10 & k\ell op=24\\
(o-1)(o-2)(o-3)(o-4)=0 & & \\
(p-1)(p-2)(p-3)(p-4)=0 & &
\end{array}$$

If we have in mind a specific Shidoku puzzle (that is, a partially filled in square), then we would need to add one further equation for each of the initial clues. For example, given the specific six-clue Shidoku puzzle below left, we would add the six very simple equations shown below right:

			4
4		2	
	3		1
1			

$$d=4 \qquad e=4 \qquad g=2$$

$$j=3 \qquad \ell=1 \qquad m=1$$

Since this Shidoku puzzle has a unique solution, adding these six equations to our forty earlier equations yields a system with a unique solution. There is only one way of assigning values to $a, b, c, \ldots, p$ that makes all forty-six equations true simultaneously.

Clearly it would be more fun to play the Shidoku puzzle than to solve all those equations. However, the point of this was not to amplify the fun; it was to produce an algebraic representation of Shidoku, in the same way that earlier we used graphs to obtain a visual representation. It is messy, but by relocating to the world of algebra and polynomials, we make possible the use of some powerful mathematical machinery.

8.2 THE PERILS OF GENERALIZATION

As clever as that is, our interest is in Sudoku, not Shidoku. That does not seem like much of a hurdle, however, since to a casual glance it would seem that everything we have done can be adjusted easily to the 9×9 case. We would begin with eighty-one individual equations telling us that each cell takes on a value from 1–9. Then for each of the twenty-seven Sudoku regions (nine rows, nine columns, and nine 3×3 blocks), we have two additional equations asserting that the relevant cells sum to $1+2+3+4+5+6+7+8+9 = 45$ and multiply to $1 \cdot 2 \cdot 3 \cdot 4 \cdot 5 \cdot 6 \cdot 7 \cdot 8 \cdot 9 = 362{,}880$. That gives a total of $81 + 54 = 135$ equations that together capture everything we need to know about our Sudoku square.

This is a situation mathematicians confront regularly. Having worked out a few concrete cases, we try to extend our examples back to the general question. It is part of our training that, when attempting such a feat, we must be ever alert that nothing relevant to our concrete case changes in returning to the original problem. In this case, alas, our work in the last paragraph overlooked a nasty point.

The Shidoku equations $w + x + y + x = 10$ and $wxyz = 24$ worked because the only way of choosing four numbers from among 1 to 4 that sum to 10 and multiply to 24 is to select each digit exactly once. If we now let $x_1, x_2, \ldots, x_9$ represent the nine cells in the same Sudoku region, then we have the comparable equations

$$x_1 + x_2 + \cdots + x_8 + x_9 = 45 \quad \text{and} \quad x_1 x_2 \ldots x_8 x_9 = 362{,}880.$$

Unfortunately, it turns out that there are *two* ways of selecting nine numbers from among 1 to 9 that satisfy both equations. There is the desired solution of choosing $x_1, x_2, \ldots, x_9$ to be the numbers $1, 2, \ldots, 9$ in some order. But there is another possibility:

Puzzle 64: Sum 45 and Product 362,880.

Try to find integers A, B, C, D, E, F, G, H, and I from 1–9 with at least one repeated value so that $A + B + C + D + E + F + G + H + I = 45$ and $ABCDEFGHI = 362{,}880$. The answer is in the text below, so don't peek until you are ready to give up!

One way of attacking this puzzle is to look at the prime factorization of $362{,}880 = 2^7 \cdot 3^4 \cdot 5^1 \cdot 7^1$ and experiment with ways of breaking that number into nine factors whose sum is 45. Or you could use some computer power. Here is the answer for when you are ready to admit defeat (or check your success): a second way to get a sum of 45 and a product of 362,880 uses the numbers 1, 2, 4, 4, 4, 5, 7, 9, 9, in some order. In other words, $1 + 2 + 4 + 4 + 4 + 5 + 7 + 9 + 9 = 45$ and $1 \cdot 2 \cdot 4 \cdot 4 \cdot 4 \cdot 5 \cdot 7 \cdot 9 \cdot 9 = 362{,}880$. Permit us to honor this solution with a puzzle:

Puzzle 65: Bad News Sudoku.

Fill in the grid so that each row, column, and block contains exactly one each of 1, 2, 5, and 7; exactly two 9s; and exactly three 4s.

		9	2				4	4
	9	7						9
1	2				4	4		
5					4	4		
				2				
		4	4					1
		4	4				7	5
9						1	2	
9	7				9	4		

The existence of this second solution is disappointing. It implies that we cannot generalize our Shidoku equations directly into a Sudoku system in the obvious way. Our sum and product equations no longer ensure that the nine cells in the same region have distinct values.

There is no need to panic, however. We could certainly add more polynomials to render invalid our alternate set of of numbers. Actually, though, we can be a bit more clever than that; instead of changing the system of polynomials, let us change the numbers themselves! As we have emphasized throughout, there is nothing special about the digits 1–9. Sudoku only requires nine different symbols, not those digits specifically. If we could find nine digits with the property that the only way to obtain their sum and product is to select each number exactly once then we would be back on track.

With the aid of a computer, it is not hard to find sets having this property. It turns out (see [6]) that among all such sets of numbers, the smallest in magnitude is $\{-2, -1, 1, 2, 3, 4, 5, 6, 7\}$. Using these digits in place of the more familiar 1–9 leads to eighty-one equations of the form

$$(w+2)(w+1)(w-1)(w-2)(w-3)(w-4)(w-5)(w-6)(w-7)=0,$$

and fifty-four equations of the form

$$x_1 + x_2 + \cdots + x_8 + x_9 = 25 \quad \text{and} \quad x_1 x_2 \ldots x_8 x_9 = 10{,}080.$$

There is no reason we cannot play Sudoku with these numbers instead, just for variety:

Puzzle 66: Shift Sudoku.

Fill in the grid so that each row, column, and block contains exactly one each of the numbers $-2, -1, 1, 2, 3, 4, 5, 6, 7$.

	-2	4			-1	3		
		2				5		
			3	5				
2	-1			4				
	1						3	
				-2			7	6
				1	5			
		-1				4		
		6	2			-2	-1	

To recap, the point is not merely that we can construct a puzzle with these numbers. We could do that with any nine numbers or symbols that we like. The point is that with the set $\{-2, -1, 1, 2, 3, 4, 5, 6, 7\}$, the sum and product polynomials for each region perfectly encode the property we want, namely that each of the numbers is used exactly once.

8.3 COMPLEX POLYNOMIALS

There are many other ways of associating systems of polynomials to Sudoku and Shidoku squares. The path we just explored used the rows, columns, and blocks to form equations describing the requirement of having distinct values in the cells of a region. This is different from the approach we used with the Sudoku graph, where we considered cells in pairs according to whether or not they lived in the same region. Perhaps there is a way of constructing a polynomial system by considering those pairs instead of entire regions at once.

In fact, there is. It involves imaginary numbers, however, so let us provide a quick review. We will need the number i, which is defined by the equation $i = \sqrt{-1}$. This implies that $i^2 = -1$, $i^3 = -i$, and $i^4 = 1$. Put differently, we could say that i is a fourth root of the number 1, or shorten that to the more common phrase that "i is a fourth root of unity." We think you will find that the other fourth roots of unity are 1, -1, and $-i$. If any of these numbers is raised to the fourth power, the result is 1. That is all we shall need.

Our new system of polynomials begins by using the symbols ± 1 and $\pm i$ in place of the more familiar digits in our Shidoku square. This construction is used by many people, including Arnold et al. [6] and Gago-Vargas [26]. To get used to this idea, take a moment for some sample puzzles:

Puzzle 67: Complex Shidoku.

Fill in each grid so that every row, column, and block contains 1, -1, i, and $-i$ exactly once. Being Shidoku, the puzzles are pretty easy; but keeping track of all the similar-looking symbols provides some challenge!

		-i	i
			1
i			
1	-i		

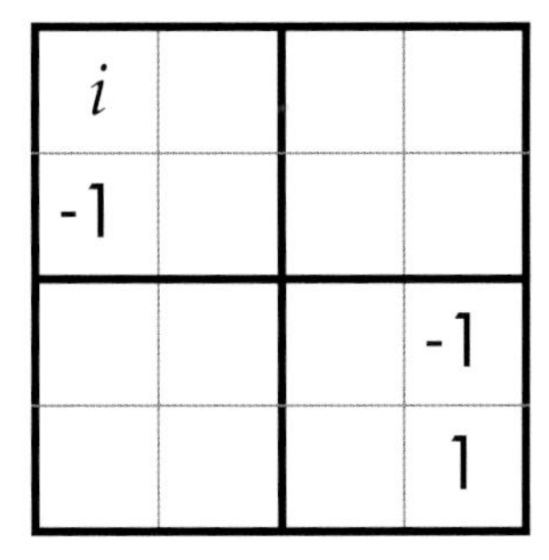

Our new requirement that each of the cells contains 1, -1, i, or $-i$ means that for each of the sixteen cells w in the puzzle we have

$$w^4 = 1.$$

The sum and product method we used earlier encoded entire regions as polynomials; every row, column, and block defined a pair of polynomials. Our new framework allows us to construct a different set of polynomials, one for each pair of cells that share a region.

To this end, let w and x denote two cells in the same region. We need an equation asserting their unequal values. We already know that the fourth power of any of $1, -1, i$, or $-i$ is equal to 1. It follows that we must have $w^4 = 1$ and $x^4 = 1$. In particular, this means that $w^4 = x^4$, and thus that

$$w^4 - x^4 = 0.$$

We can factor $w^4 - x^4$ as $(w^2 - x^2)(w^2 + x^2)$, which further factors as $(w - x)(w + x)(w^2 + x^2)$. This means that the equation above is equivalent to the equation

$$(w - x)(w + x)(w^2 + x^2) = 0.$$

The requirement that w and x take on different values says that $w \neq x$, which means that $w - x \neq 0$. Therefore, the only way that $(w - x)(w + x)(w^2 + x^2)$ could be 0 is if the second or third factor is 0. That means that our equation above is equivalent to the equation

$$(w + x)(w^2 + x^2) = 0.$$

Let us review: for each cell w in the square, we get an equation of the form $w^4 = 1$. That's sixteen equations. Also, for each pair w and x of cells that lie in the same row, column, and block, we get an equation of the form $(w + x)(w^2 + x^2) = 0$. How many of those equations will we have? Well, in each of the four rows there are six ways of choosing two elements. That gives us twenty-four equations. An additional twenty-four equations arise from the six ways of choosing two elements from each of the four columns. Each of the four blocks now contributes two additional equations (for the diagonal pairs), for a grand total of fifty-six equations.

If you think about it, these fifty-six pairs of cells are exactly the ones that were connected by our "Shidoku spiders" from the previous chapter:

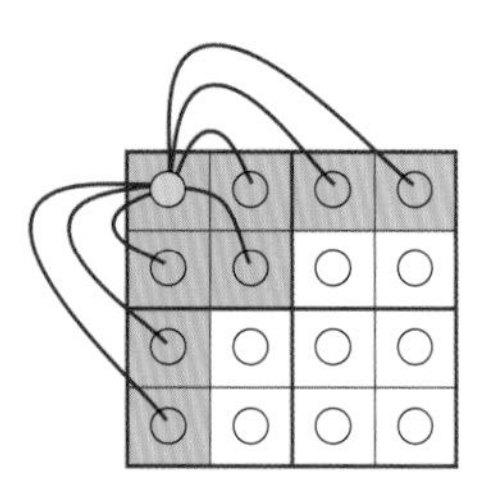

As we saw before, each cell on the board is paired with three others in its row, three others in its column, and one other in its block, for a total of seven pairs involving that cell. Since there are sixteen cells on the board and each of these pairs involves two cells, we now have another way of seeing that there are $\frac{7(16)}{2} = 56$ equations.

The fifty-six equations arising from pairs, together with the sixteen equations arising from individual cells, gives us seventy-two equations in the roots-of-unity

Shidoku system. Using the same variable names $a, b, c, \ldots, p$ as in the previous section, these seventy-two equations are:

$$\begin{array}{llll}
a^4 = 1 & (a+b)(a^2+b^2)=0 & (d+h)(d^2+h^2)=0 & (i+\ell)(i^2+\ell^2)=0 \\
b^4 = 1 & (a+c)(a^2+c^2)=0 & (d+\ell)(d^2+\ell^2)=0 & (i+m)(i^2+m^2)=0 \\
c^4 = 1 & (a+d)(a^2+d^2)=0 & (d+p)(d^2+p^2)=0 & (i+n)(i^2+n^2)=0 \\
d^4 = 1 & (a+e)(a^2+e^2)=0 & (e+f)(e^2+f^2)=0 & (j+k)(j^2+k^2)=0 \\
e^4 = 1 & (a+f)(a^2+f^2)=0 & (e+g)(e^2+g^2)=0 & (j+\ell)(j^2+\ell^2)=0 \\
f^4 = 1 & (a+i)(a^2+i^2)=0 & (e+h)(e^2+h^2)=0 & (j+m)(j^2+m^2)=0 \\
g^4 = 1 & (a+m)(a^2+m^2)=0 & (e+i)(e^2+i^2)=0 & (j+n)(j^2+n^2)=0 \\
h^4 = 1 & (b+c)(b^2+c^2)=0 & (e+m)(e^2+m^2)=0 & (k+\ell)(k^2+\ell^2)=0 \\
i^4 = 1 & (b+d)(b^2+d^2)=0 & (f+g)(f^2+g^2)=0 & (k+o)(k^2+o^2)=0 \\
j^4 = 1 & (b+e)(b^2+e^2)=0 & (f+h)(f^2+h^2)=0 & (k+p)(k^2+p^2)=0 \\
k^4 = 1 & (b+f)(b^2+f^2)=0 & (f+j)(f^2+j^2)=0 & (\ell+o)(\ell^2+o^2)=0 \\
\ell^4 = 1 & (b+j)(b^2+j^2)=0 & (f+n)(f^2+n^2)=0 & (\ell+p)(\ell^2+p^2)=0 \\
m^4 = 1 & (b+n)(b^2+n^2)=0 & (g+h)(g^2+h^2)=0 & (m+n)(m^2+n^2)=0 \\
n^4 = 1 & (c+d)(c^2+d^2)=0 & (g+k)(g^2+k^2)=0 & (m+o)(m^2+o^2)=0 \\
o^4 = 1 & (c+g)(c^2+g^2)=0 & (g+o)(g^2+o^2)=0 & (m+p)(m^2+pb^2)=0 \\
p^4 = 1 & (c+h)(c^2+h^2)=0 & (h+\ell)(h^2+\ell^2)=0 & (n+o)(n^2+o^2)=0 \\
 & (c+k)(c^2+k^2)=0 & (h+p)(h^2+p^2)=0 & (n+p)(n^2+p^2)=0 \\
 & (c+o)(c^2+o^2)=0 & (i+j)(i^2+j^2)=0 & (o+p)(o^2+p^2)=0 \\
 & (d+g)(d^2+g^2)=0 & (i+k)(i^2+k^2)=0 &
\end{array}$$

These seventy-two equations perfectly define the structure of a Shidoku square. We now have a triangle of equivalences. A given Shidoku board, a properly four-colored Shidoku graph, and a solution to our system of seventy-two complex equations are all really the same thing in different languages.

This new system of polynomials is larger than our previous one and involves complex numbers, but computationally it may be simpler than our previous system. The roots-of-unity system is really no more or less unwieldy than our sums-and-products system. It has the advantage of generalizing directly to the 9×9 case (in which case we would use ninth roots of unity, of course). We can use computers to find solutions to either type of system. In the 4×4 cases this does not require much computing time, but in the 9×9 case even modern computing power is inadequate. What is needed is a tool for simplifying our systems of equations without losing any essential information. Modern algebra provides just such a tool, known as a *Gröbner basis*. With some regret we have decided such methods are too complex to describe, in a reasonable amount of space, in this book. We encourage you to look at the paper by Arnold et al. [6] for this and other algebraic approaches to the problem.

8.4 THE RISE OF EXPERIMENTAL MATHEMATICS

The computer has been our constant companion throughout this discussion. Human ingenuity is fine when you are trying to devise clever systems of polynomials, but when it comes time to solve them you need a mechanical assist. We have seen this sort of thing before. Our count of the number of Sudoku squares likewise involved a mix of ingenuity and hardware. In our brief discussion of the Four-Color Theorem we saw the same thing, and so it is again here. Indeed, the problem of using computers to solve and create Sudoku puzzles was one of the original motivations for thinking along the lines described in this chapter. (Consult the paper by Bartlett et al. [9] for a detailed discussion of this point.)

The ready availability of prodigious computing power has returned to prominence a long-hidden aspect of mathematics. We are referring to experimentation. The superprecise definitions and rigorous proofs of the textbooks are the end result of mathematical investigation. They are not the sum total of it. It has been wittily observed that a mathematician is a machine for turning coffee into theorems. There is truth in this, so long as we understand the precise role of the coffee. It is to fuel the long hours of trial and error and messing around that precede the emergence of a polished result. You must work out many examples and special cases before catching a glimpse of what is true.

This sort of experimentation has been a central part of mathematical practice for as long as people have engaged in the enterprise. The computer's role is to increase dramatically the number and complexity of examples that can be worked out. As a simple example, consider Goldbach's conjecture. It asserts that every even number greater than 2 can be expressed as the sum of two primes (possibly in more than one way). We can certainly check the assertion by hand for small even numbers. We have that

$$4 = 2 + 2 \qquad 6 = 3 + 3 \qquad 8 = 3 + 5 \qquad 10 = 3 + 7 \qquad 12 = 5 + 7,$$

for example. This becomes tedious in a big hurry. Even if we work very quickly, there is still a limit to how many examples we could check in this way. The computer is far less limited. With its aid, Goldbach's conjecture has been checked for all even numbers up to 10^{18} with no counterexample being found.

We now have two examples of how the computer is used in mathematical practice. It can produce proofs by exhaustion by checking large numbers of cases, as in our count of the Sudoku squares or in the proof of the four-color theorem. It can also search for counterexamples to conjectures. On top of this, computers have also been used to discover obscure patterns among numbers and other mathematical objects and to evaluate infinite sums and products. Sophisticated graphics packages have made it possible to visualize relationships among abstract constructions. All of these techniques, and others besides, have been used to make progress in mathematical research. (The excellent book by Borwein and Devlin [13] provides numerous concrete examples).

Things have developed to the point where experimental mathematics is now a fully established branch of the discipline, complete with its own research journals and textbooks. The classical sort of mathematics still dominates, certainly, and deductive proof remains the unshakeable gold standard of mathematical truth. It is just that, increasingly, human ingenuity is being directed not toward producing traditional proofs, but toward reducing difficult problems to computationally tractable forms. Devising clever experiments for a computer to carry out is today a fully legitimate activity for professional mathematicians.

All of which are useful reminders that mathematics is not a static discipline. Both the problems that are studied and the standards for determining that a problem has been solved change with the times. Which raises some interesting questions. In the interest of leaving you with some food for thought, consider the following paragraph from mathematician Keith Devlin [18]. You will need to know that the Riemann Hypothesis is a conjecture regarding the location of the zeroes of a particular function.

> *The degree to which mathematics has come to resemble the natural sciences can be illustrated using the example I have already cited: the Riemann Hypothesis. As I mentioned, the hypothesis has been verified computationally for the ten trillion zeros closest to the origin. But every mathematician will agree that this does not amount to a conclusive proof. Now suppose that, next week, a mathematician posts on the Internet a five-hundred page argument that she or he claims is a proof of the hypothesis. The argument is very dense and contains several new and very deep ideas. Several years go by, during which many mathematicians around the world pore over the proof in every detail, and although they discover (and continue to discover) errors, in each case they or someone else (including the original author) is able to find a correction. At what point does the mathematical community as a whole declare that the hypothesis has indeed been proved? And even then, which do you find more convincing, the fact that there is an argument – which you have never read, and have no intention of reading – for which none of the hundred or so errors found so far have proved to be fatal, or the fact that the hypothesis has been verified computationally (and, we shall assume, with total certainty) for 10 trillion cases?*

9

Extremes

Sudoku Pushed to Its Limits

Mathematicians have a reputation for being a staid and placid lot, but that is really an unfair characterization. You see, in our professional lives, we are constantly taking things to extremes. We are endlessly pushing the envelope as far it can go, and even then we are not satisfied until we have actually proven that it can go no farther. We do this because you cannot properly understand a concept until you have established its boundaries. It is as though our abstractions were long rubber walls blocking our view of what lies beyond them. By pushing on the rubber, we expose some of the ground beneath. The more we push, the more ground we reveal. Exposing the maximal amount of ground means pushing on the rubber until it breaks.

For that reason, we devote the present chapter to a gallery of Sudoku extremes. In most cases, we cannot yet prove that we have pushed things to their utmost. That means there is an implicit challenge running through most of our examples. This is the best *we* can do. Can *you* do any better?

9.1 THE JOYS OF GOING TO EXTREMES

In many cases, there is practical benefit in going to extremes. In calculus, we spend a great deal of time finding maximum and minimum points of functions, because those extremal points often represent important physical quantities. Such points could represent the highest altitude attained by a projectile, for instance, or the combination of expenditures likely to maximize future profit. Our interest goes beyond this, however. Searching for the weird and the unusual, the *extreme*, encourages habits of mind unsatisfied with superficial appearances. Such examples make you realize that "intuitively satisfying" is not synonymous with "true." Permit us an example.

Mathematicians routinely talk about *sets*, by which we simply mean collections of objects. Referring to "the set of natural numbers," for example, provides a useful formalism for discussing the entirety of the natural numbers as a single entity, as opposed to as a collection of infinitely many individual objects.

If there is one thing that is just dead bang obvious about sets, it is that for any description we give we should be able to consider the set of all objects satisfying that description. I can talk about "the set of all green things," or "the set of all apples at the local market," or "the set of all books in my office." In each case, it would seem that I have a properly defined set. Hand me any object and I can quickly check whether it does, or does not, meet the criterion.

This was "obvious," even to mathematicians, until Bertrand Russell thought to define the set S by the description "the set of all sets that are not elements of themselves." His idea was that some sets satisfy their own descriptions. "The set of all abstract ideas" is an abstract idea. It is therefore an element of itself. Likewise for "the set of all sets containing more than three elements." On the other hand, most sets are not elements of themselves. "The set of all students in my calculus class," is not itself a student in my calculus class. So we will place the sets that are elements of themselves *over there* and gather all of the others *over here*. The ones over here are the elements of S.

We now ask a simple question. Is S an element of itself? If we have a properly defined set, then we should be able to answer this question. Let us suppose S is an element of itself. Then S must satisfy its own description. That description says that S is not an element of itself, which is a contradiction. The implication is that S is not an element of itself. In this case it does, indeed, satisfy the description that defines the set S. It seems S is an element of itself after all.

Avoiding the force of this paradox required establishing elaborate axioms for set theory. Basically, mathematicians had to write down formal rules that said, in effect, "Do not be too clever in defining your sets!" If even the idea of defining a set, which must surely be among the simplest things we do as mathematicians, leads to serious problems when pushed too far, then truly there is nothing that can be taken only at face value.

9.2 MAXIMAL NUMBERS OF CLUES

We begin our discussion of Sudoku extremes with the following question: What is the maximal number of starting clues a Sudoku puzzle can have? Well, that's easy; the answer is eighty-one:

Puzzle 68: Sudoku for Busy People.

Fill in the grid so that each row, column, and block contains 1–9 exactly once.

7	3	5	6	1	2	8	9	4
6	4	9	3	8	5	1	2	7
1	2	8	4	7	9	3	5	6
2	5	1	9	6	3	7	4	8
4	9	6	8	2	7	5	3	1
8	7	3	1	5	4	2	6	9
9	8	4	2	3	1	6	7	5
5	1	2	7	4	6	9	8	3
3	6	7	5	9	8	4	1	2

That was very easy, meaning it was not such an interesting question. Here is one that is slightly more interesting: What is the maximal number of starting clues a puzzle can have without having a unique solution?

As it turns out, that one is easy too. The answer is seventy-seven. It is easy to see that if you have seventy-eight, seventy-nine, or eighty starting clues then the puzzle can only be completed in one way. But with seventy-seven clues we have the possibility of omitting a toggle. For example, consider these seventy-seven clues from the "busy people" Sudoku puzzle above:

This is not a puzzle

7	3	5	6	1	2			4
6	4	9	3	8	5	1	2	7
1	2	8	4	7	9	3	5	6
2	5	1	9	6	3	7	4	8
4	9	6	8	2	7	5	3	1
8	7	3	1	5	4	2	6	9
9	8	4	2	3	1	6	7	5
5	1	2	7	4	6			3
3	6	7	5	9	8	4	1	2

Even with so many clues, this seventy-seven-clue pseudo-puzzle does not have a unique solution; there is a toggle that allows two different solutions. One is the square from Puzzle 68, while the other is the same square with the 8s and 9s reversed in the green cells:

7	3	5	6	1	2	8	9	4
6	4	9	3	8	5	1	2	7
1	2	8	4	7	9	3	5	6
2	5	1	9	6	3	7	4	8
4	9	6	8	2	7	5	3	1
8	7	3	1	5	4	2	6	9
9	8	4	2	3	1	6	7	5
5	1	2	7	4	6	9	8	3
3	6	7	5	9	8	4	1	2

7	3	5	6	1	2	9	8	4
6	4	9	3	8	5	1	2	7
1	2	8	4	7	9	3	5	6
2	5	1	9	6	3	7	4	8
4	9	6	8	2	7	5	3	1
8	7	3	1	5	4	2	6	9
9	8	4	2	3	1	6	7	5
5	1	2	7	4	6	8	9	3
3	6	7	5	9	8	4	1	2

In fact, for a Sudoku square of size $n \times n$, the maximal number of clues that does not guarantee a unique solution is $n^2 - 4$; a toggle is the smallest ambiguity we can have.

Let us move on to a more challenging question. What is the maximum number of *independent* starting clues that a Sudoku puzzle can have? In other words, what is the maximal number of clues such that the removal of *any* one of them renders the puzzle unsound. This question is sufficiently difficult that we do not know the answer. Let us start with a puzzle:

Puzzle 69: More-Than-You-Need Sudoku.

Fill in the grid so that each row, column, and block contains 1–9 exactly once. The solution to this puzzle is the Sudoku square we were investigating previously, so don't look back!

7			6	1	2			
	4	9						
1			4			3		
	5	1						
	9		8		7		3	
						2	6	
		4			1			5
						9	8	
			5	9	8			2

This puzzle has twenty-six clues and is of easy-to-moderate difficulty. Not all the clues are necessary, however. There are three clues that could each be individually removed without affecting the puzzle's soundness. They are shown below in green:

7			6	1	2			
	4	9						
1			4			3		
	5	1						
	9		8		7		3	
						2	6	
		4			1			5
						9	8	
			5	9	8			2

Mind you, we cannot remove all three at once; that would leave a pseudo-puzzle with five solutions. But any individual green clue can be removed to leave us with a sound 25-clue puzzle. In fact, we can even get two sound 24-clue puzzles; one by removing both the 1 and the 7, and the other by removing the 1 and the 9:

7			6	1	2			
	4	9						
			4			3		
	5	1						
	9		8				3	
						2	6	
		4			1			5
						9	8	
			5	9	8			2

7			6	1	2			
	4	9						
			4			3		
	5	1						
			8		7		3	
						2	6	
		4			1			5
						9	8	
			5	9	8			2

It turns out that removing both the 7 and the 9 leaves an unsound puzzle, if you are curious. The two 24-clue puzzles are more difficult than the original 26-clue puzzle, but still only moderately so. The three nonrequired clues were included in the original to round out the rotational symmetry and also to make the puzzle easy enough to play. The vast majority of Sudoku puzzles found in books and newspapers contain extraneous clues.

What have we answered here? We found two 24-clue puzzles that cannot be made any smaller. Each clue in these puzzles is essential and plays a part in getting us to the unique solution square. We say that these 24-clue puzzles are *irreducible.* Adding clues to either of our 24-clue irreducible puzzles results in a reducible puzzle. We have found a lower bound for the maximal number of independent starting clues. The maximum number of independent clues must be greater than or equal to twenty-four.

Can we do better? Well, neither of our 24-clue irreducible puzzles had the usual 180-degree rotational symmetry. Is it possible to find a puzzle on this Sudoku square that has twenty-four independent clues and also has this symmetry? It turns out that we can:

Puzzle 70: Just-What-I-Needed Sudoku.

Fill in the grid so that each row, column, and block contains 1–9 exactly once. Again, the solution to this puzzle is the Sudoku square we were just investigating, so no peeking.

				1	2			4
	4					1	2	
		8		7			5	
			9				4	
		6				5		
	7				4			
	8			3		6		
	1	2					8	
3			5	9				

Can we do even better? Could there be a 25-clue irreducible puzzle on this Sudoku square? Or one with an even greater number of independent clues? The answer is yes, but if we want the largest number possible then we will have to drop the symmetry condition. The current known maximum for any puzzle is thirty-nine independent clues. Here is such a puzzle, on our now-familiar solution square, that is based on a puzzle from the Sudoku Programmers Forum [41]:

Puzzle 71: Maximum Independent Sudoku.

Fill in the grid so that each row, column, and block contains 1–9 exactly once. The solution to this puzzle is once again the Sudoku square we have been investigating.

7		5	6			8		4
6	4						2	7
1	2	8	4	7			5	6
2	5	1		6				8
8				5		2	6	
	8			3			7	
5		2	7	4			8	3
3		7	5			4		2

Can you do forty? Currently there is no proof to show that you cannot. The best we can say is that there exist instances of 39-clue irreducible puzzles, but there are no known instances of 40-clue irreducible puzzles.

We can give definitive answers to these questions in the easier 4×4 case of Shidoku. The maximal number of independent clues is known to be six in this case, as can be established by an exhaustive computer search. How much calculation is required to establish this result?

We must first check that there are no seven-clue irreducible puzzles. In Section 5.3 we saw that there are only two essentially different types of Shidoku squares, and we had these representatives for each type:

Type 1 Shidoku

1	2	3	4
3	4	1	2
2	1	4	3
4	3	2	1

Type 2 Shidoku

1	2	3	4
3	4	1	2
2	3	4	1
4	1	2	3

Counting the number of seven-clue subsets for one of these boards is a standard problem in combinatorics. The answer is given by $\binom{16}{7} = \frac{16!}{7!\,9!} = 11{,}440$. This number must be doubled to 22,880 to account for the two types of squares.

Some ways of selecting seven clues do not produce sound puzzles, and they can be discarded. Among those that remain, we systematically remove one clue from each in all possible ways. If we are correct that there are no seven-clue irreducible puzzles, then there is at least one clue on each board that can be removed without harming the puzzle's soundness.

These numbers are not really so large, which is why an exhaustive computer search is possible. The result is a demonstration that the maximal number of clues is no more than six. Showing the number is exactly six requires producing a six-clue irreducible puzzle. Here is such an example:

	2	3	
3		1	
2	1		

Interestingly, simple enumeration by computer shows that only Type 1 squares have six-clue irreducible puzzles. The maximum number of independent clues in a puzzle for a Type 2 square is five.

As we have noted before, though the computer proof is nice, a classical proof would be better. In particular, a nonenumerative proof in the Shidoku case might provide a useful clue for establishing the maximum number in the computationally intractable Sudoku case. The enumerative argument for 4×4 Shidoku is unhelpful in the larger 9×9 Sudoku universe. We are certainly not prepared to look at all $\binom{81}{40} = \frac{81!}{40!\,41!} = 212{,}392{,}290{,}424{,}395{,}860{,}814{,}420$ of the forty-clue subsets of each of the possible 5,472,730,538 types of Sudoku boards (see Section 5.7), even with the help of a hefty computer.

Although there is as yet no elegant argument to show that a Shidoku puzzle could never have seven independent clues, we can establish an upper bound by classical means. Specifically, we can show that no Shidoku puzzle can ever have nine independent clues.

Remember that we only have to make the argument for one representative of a Type 1 Shidoku square and one representative of a Type 2 Shidoku square. Here are such representatives with a certain coloring which we shall explain in a moment:

Type 1 Shidoku

1	2	3	4
3	4	1	2
2	1	4	3
4	3	2	1

Type 2 Shidoku

1	2	3	4
3	4	1	2
2	3	4	1
4	1	2	3

We will present the argument for the Type 1 board, and leave the (nearly identical) Type 2 case as an exercise for you. Above we have divided the Type 1 board into four (disconnected) regions, colored orange, green, purple, and yellow.

We claim that in any puzzle on this board, each of the four regions can contain at most two independent clues. Since the four regions are disjoint, this implies there can be at most eight independent clues in any puzzle for this board.

Our proof hinges on a very simple claim. If any three of the cells in the same colored region are filled in, then it must be the case that one of them is actually implied by the other two. For example, on the Type 1 board, there are four ways of choosing three clues in the orange region:

1	2	3	4
3	4	1	2
2	1	4	3
4	3	2	1

1	2	3	4
3	4	1	2
2	1	4	3
4	3	2	1

1	2	3	4
3	4	1	2
2	1	4	3
4	3	2	1

1	2	3	4
3	4	1	2
2	1	4	3
4	3	2	1

In each case, it is apparent that the Shidoku rules force the triangle clue to be a consequence of the two circle clues. Notice that we are not saying just any two clues in the orange region automatically imply a given third clue. The claim, rather, is that any three clues in the orange region must contain two that imply the third. You have to choose carefully, but once you do, the argument is straightforward. The same argument applies to the green, purple, and yellow regions. It follows that there is no way a puzzle on either board could have nine independent clues.

Can you find an argument that gets us down to six, that is, that shows that seven independent clues is impossible? If you can, then you will have moved the boundary of mathematics forward another tiny inch. Let us know if you do!

9.3 THREE AMUSING EXTREMES

Extreme examples can be interesting simply for their manifest deviation from expected norms. If you browse the puzzles in a typical Sudoku magazine you will find the starting clues distributed roughly evenly across the square. But do they *have* to be so? How far can we push the envelope?

For example, if you came across a puzzle like the one below, you would immediately notice that three entire blocks of the puzzle have no clues at all.

Puzzle 72: Off-Diagonal Sudoku.

Fill in the grid so that each row, column, and block contains 1–9 exactly once. Note that there are no clues in any of the three blocks along the main diagonal.

			1			9	6	
							1	4
			9		8	7		2
1		6				8		
		7				3		6
8		2	6		9			
4	9							
	3	1			2			

Even more striking would be a puzzle that missed a huge chunk of cells in the middle of the board:

Puzzle 73: Empty Space Sudoku.

Fill in the grid so that each row, column, and block contains 1–9 exactly once. Note that there are no clues in the large 5 × 5 square in the center.

	6		5					
	9	3	8	1	2	5	6	4
	5						8	
	3						9	8
	1						4	
2	4						3	
	7						5	
1	8	9	3	4	5	2	7	
					6		1	

Computer scientists and mathematicians enjoy searching for such geometrically extreme puzzles. Currently, the largest known block of cells that can be missed is 5 × 6, which is slightly bigger than the one in our example. (Our puzzle, on the other hand, satisfies our zeal for rotational symmetry.) How can we prove that 5 × 6 is the largest rectangle we can miss? We do not know. Yet. For now, we will just enjoy the puzzles.

What is the maximum number of regions that can be vacant in a sound puzzle? The answer is currently thought to be nine. Here is an example in which three blocks, three rows, and three columns are all empty. In addition, it also misses the 5 × 5 square in the center of the board.

Puzzle 74: Avoidance Sudoku.

Fill in the grid so that each row, column, and block contains 1–9 exactly once. Note that nine regions, as well as the 5×5 center group of cells, contain no initial starting clues.

			4		8		1	2
			9		3		5	7
4	9						7	1
8	6						3	5
7	4		1		9			
6	8		2		4			

9.4 THE ROCK STAR PROBLEM

Which brings us to the most important extreme of them all. What is the smallest number of starting clues needed for a sound Sudoku puzzle? This question remains open. If you can figure it out, you will be a rock star in the universe of people who care about such things. Granted, that is a far smaller universe than the one full of people who care about actual rock stars, but still, it would be great.

In the hopes that an industrious reader will be the one to break this problem, we present it formally:

Open Question: *What is the minimum number of starting clues possible in a Sudoku puzzle with a unique solution?*

We do know a few things. For example, we have already seen that we need at least eight starting clues in any sound puzzle, since we need to use at least eight of the nine numbers to guarantee a unique solution. On the other hand, we have already seen several examples of eighteen-clue Sudoku. The answer to this problem, then, is certainly between eight and eighteen inclusive.

In fact, we can do better. Earlier we saw that finding an 18-clue rotationally symmetric Sudoku puzzle is akin to finding a needle in a haystack. Finding 17-clue Sudoku puzzles is the same, though many are known. As of the writing (in April 2011), Gordon Royle [35] has compiled a list of over 49,000 essentially different 17-clue Sudoku puzzles. Here is one of them (number 4,200 from the list, in celebration of the universally important number 42 from Adams [1]):

Puzzle 75: 17-Clue Sudoku.

Fill in the grid so that each row, column, and block contains 1–9 exactly once.

				4			2	
	5		9					
	1							
			8			1		5
2				3				
						9		
4	9				2			
3							6	
			1					

Royle has checked that none of the seventeen-clue puzzles in his list are reducible. Each clue in every puzzle is independent of the others and cannot be removed. That means no sixteen-clue puzzle is hiding inside any of these seventeen-clue puzzles. In fact, nobody has ever found an example of a sound sixteen-clue puzzle. Alas, examining fifty thousand pieces of hay without finding a needle does not imply there is no needle to be found. There is no mathematical proof – computational or otherwise – that seventeen is the minimum number.

The haystack is just too big for a direct computer search. In order to know definitively that absolutely no 16-clue puzzles exist, we would have to check every 16-clue subset of the 5,472,730,538 essentially different Sudoku squares (see Chapter 5). That's

$$183,851,407,423,359,414,572,057,730$$

different 16-clue candidates we are talking about. We simply cannot do that with current computing power.

Of course, we could get lucky and stumble across a sixteen-clue puzzle after just a few attempts. But if no such puzzle exists then we would have to consider all of the possibilities to be sure. So the open question boils down to this:

Open Question: *Does a sound sixteen-clue Sudoku puzzle exist?*

You may notice that the seventeen-clue grid in Puzzle 75 does not have our preferred rotational symmetry. That is because there are *no* known examples of rotationally symmetric seventeen-clue puzzles. This suggests a second question:

Open Question: *Does a sound, rotationally-symmetric seventeen-clue Sudoku puzzle exist?*

If we add the requirement of symmetry to the clue placement, our conjectured minimum number rises from seventeen to eighteen. We can also go the other way: if we add conditions to the board itself, we can decrease the number of clues that are needed. To this end, Ruud van der Werf [37] has compiled a list of over seven thousand twelve-clue Sudoku X puzzles. That is, puzzles with the added condition that the two main diagonals are also Sudoku regions. Here is one of them:

Puzzle 76: Twelve-Clue Sudoku X.

Fill in the grid so that each row, column, block, and main diagonal contains 1–9 exactly once. Sit down in a comfortable seat; this one is tough!

							1	
						2		
	3					4		5
					6			
				7				
6		2					8	
			3	4				

You can probably guess what comes next: No eleven-clue Sudoku X puzzle has ever been found. But once again, that is not a proof; just the seed for a new open question:

Open Question: *Does a sound eleven-clue Sudoku X puzzle exist?*

One way of tackling these sorts of minimum-clue Sudoku problems is to consider the so-called *unavoidable sets* mentioned in Chapter 6. Recall that unavoidable sets are subsets of Sudoku squares that must be hit by the clues of any sound puzzle. To refresh our memories, let us consider this concept in the case of 4×4 Shidoku. It will turn out that this line of attack actually solves the Rock Star problem in this smaller case.

We shall prove that the minimal number of clues in a sound Shidoku puzzle is four. The proof has two steps. First we give an example of a sound, four-clue Shidoku puzzle. Then we will use unavoidable sets to prove that a sound three-clue puzzle is not possible.

Begin by playing the following Shidoku puzzle. You will find it can be completed in only one way:

1			
		3	
	2		
			4

That establishes the possibility of a sound four-clue puzzle. Now we must show there are no sound three-clue puzzles. We shall follow the method given in Taalman [43]. Recall that there are only two fundamentally different Shidoku squares, colored here in a useful manner:

Type 1 Shidoku

1	2	3	4
3	4	1	2
2	1	4	3
4	3	2	1

Type 2 Shidoku

1	2	3	4
3	4	1	2
2	3	4	1
4	1	2	3

The four colored regions in each of these squares are unavoidable sets. Let us review what that means on the Type 1 square: Even if we knew every entry in the square except for the ones in the yellow cells, it would still be impossible to determine the values in those cells:

	2		4
	4		2
2	1	4	3
4	3	2	1

It follows that every Shidoku puzzle whose solution is the Type 1 square *must* contain at least one clue in the yellow set. Thus, the four yellow cells form an unavoidable set.

The other colored regions have the same property, and none of the four unavoidable sets overlap. This is true in the Type 2 square as well. Since we must have at least one starting clue from each of four nonoverlapping sets, we see that a sound Shidoku puzzle must have at least four starting clues.

It is nice that a complete solution can be given in the Shidoku case, but our argument does not extend to a Sudoku argument. There are two reasons for this. First, in the 4×4 case there are only two essentially different types of squares, while in the 9×9 case there are over five billion. Worse, in the Shidoku case, it is clear how to split each type of board into nonoverlapping unavoidable sets. In the Sudoku case, it is unclear how to identify seventeen nonoverlapping unavoidable sets.

Worse still, though both types of Shidoku squares admit four-clue puzzles, there are many Sudoku squares that do not admit seventeen-clue puzzles. That is, there are Sudoku squares with the property that every puzzle with that square for a solution must have more than seventeen clues. Large amounts of structure in the square itself can lead to many toggle-like situations. For example, consider the following square:

1	2	3	4	5	6	7	8	9
7	8	9	1	2	3	4	5	6
4	5	6	7	8	9	1	2	3
3	1	2	6	4	5	9	7	8
9	7	8	3	1	2	6	4	5
6	4	5	9	7	8	3	1	2
2	3	1	5	6	4	8	9	7
8	9	7	2	3	1	5	6	4
5	6	4	8	9	7	2	3	1

There are nine colored regions highlighted here. Each is a sort of triple-toggle. For example, consider what happens if we knew the values of all cells in the square except for those in the upper blue region:

	2	3		5	6		8	9
	8	9		2	3		5	6
	5	6		8	9		2	3
3	1	2	6	4	5	9	7	8
9	7	8	3	1	2	6	4	5
6	4	5	9	7	8	3	1	2
2	3	1	5	6	4	8	9	7
8	9	7	2	3	1	5	6	4
5	6	4	8	9	7	2	3	1

Without a clue from this region, we cannot hope to find a unique solution. The blue region is an unavoidable set. In fact, we can say more: we need *two* clues from each region. Even if we place a 1 in the uppermost left corner, say, we could still switch the 7s and 4s. The nine colored regions in our Sudoku square are unavoidable sets requiring two clues to unlock. That means that any puzzle whose solution is this Sudoku square must have at least eighteen clues.

Just for fun, we could think about a sort of inverse to this problem. Instead of looking for a Sudoku square that has no seventeen-clue puzzles, what if we look for a board that has as many seventeen-clue puzzles as possible? The current record holder is this square, found by Gordon Royle:

The Strangely Familiar Sudoku Square

6	3	9	2	4	1	7	8	5
2	8	4	7	6	5	1	9	3
5	1	7	9	8	3	6	2	4
1	2	3	8	5	7	9	4	6
7	9	6	4	3	2	8	5	1
4	5	8	6	1	9	2	3	7
3	4	2	1	7	8	5	6	9
8	6	1	5	9	4	3	7	2
9	7	5	3	2	6	4	1	8

There are twenty-nine seventeen-clue Sudoku puzzles with this square as their solution! This is known by some Sudoku explorers as the *Strangely Familiar* square because many people have examined it at length in their search for a sixteen-clue puzzle. One might hope that such a square would be more likely than most to contain a sixteen-clue puzzle. Alas, according to McGuire [29], an exhaustive computer search of all $\binom{81}{16} = 33{,}594{,}090{,}947{,}249{,}085$ of the sixteen-clue subsets of this board has shown that no sixteen-clue puzzle has this square as a solution.

So what do you think? Is seventeen the fewest clues we can have? Or is there a sixteen-clue puzzle lurking out there?

9.5 IS THERE "EVIDENCE" IN MATHEMATICS?

Within the community of mathematicians interested in Sudoku-related questions, it is near universally believed that seventeen is the minimal number of clues in a sound puzzle. Our inability to find a sixteen-clue puzzle after so much searching seems like a strong argument for believing that such a thing does not exist. Chapter 8 provided two other examples of this sort of thinking. Most mathematicians would say they believe both the Riemann Hypothesis and Goldbach Conjecture based on the extensive experimental evidence in their favor. We even suggested that concrete data can seem more convincing than a highly complex formal proof that is beyond our ability to check.

Does this not strike you as odd? In mathematics, we traffic in logical certainty. We *prove* things. There is no "pretty sure" or "highly confident." This is often presented as a selling point for our discipline as compared to our less-fortunate colleagues in the sciences. They are the ones who risk watching cherished theories collapse under the weight of new evidence. In mathematics, we have the satisfaction of knowing that when we have proved something, it stays proved.

This difference is well-captured in the distinction between deductive and inductive reasoning. The word *deduction* comes from Latin words that translate roughly as "to lead down from." The idea is that we begin with certain postulates or axioms and apply the principles of logic to derive consequences from them. *Induction* comes from Latin words meaning "to lead into." In practice, this means that we use the accumulated evidence of many specific instances to lead us into a general conclusion.

Scientists rely largely on inductive reasoning. They test their theories against the empirical data of the physical world. With each test a theory passes their confidence in its correctness increases. Our best theories have passed so many experimental tests that we refer to them as true, but we must be clear that nothing in science is ever established beyond all doubt. We tolerate this level of uncertainty because there is no alternative. In trying to understand the workings of nature, what else can we do but stick with what works until we have some reason for discarding it? We must allow the evidence of numerous concrete instances to lead us into general conclusions.

Mathematics does not study the natural world, at least not directly. We study abstract models that are sometimes based on real-world considerations, but certainly do not have to be. Our abstract models exist in a realm whose rules are under our control. We reason down from these rules to general truths about the realm we have constructed. It is all very tidy.

So why does modern mathematical practice now include such a prominent role for experimentation?

The answer, we believe, is that it is an error to view this role as any great change from tradition. Mathematics has always had a place of honor for messing around and trial and error. Just as walking precedes running, so too does groping in the dark precede a polished proof. The computer allows us to mess around in more sophisticated ways than was previously possible.

While the computer is invaluable for suggesting fruitful lines of research, we would stress that logical proof remains the gold standard. In casual conversation, we might affirm our confidence in the Goldbach Conjecture, or the Riemann Hypothesis, or the seventeen-clue conjecture, but in practice such affirmations mean nothing at all. It is just a convenient way of speaking, for how else are we to describe what we have learned from our experiments, except by using the language of evidence and confidence?

We are wading into difficult philosophical waters here. If you want a more detailed treatment of these questions from the perspective of a philosopher of mathematics we recommend the paper by Baker [8]. We also recommend the brief essay by Devlin [19] for examples where suggestive numerical evidence nonetheless leads us down the wrong path.

For our purposes, though, it is enough to note that Sudoku does not just lead naturally into topics in higher mathematics, but into the philosophy of mathematics as well.

9.6 SUDOKU IS MATH IN THE SMALL

We have come to the end of our explorations. The existing literature on the mathematics of Sudoku is far larger than we have discussed here, and it will surely grow much larger while this book is in production.

That notwithstanding, we hope our point is made. Sudoku is math in the small. If you enjoy solving Sudoku puzzles then you enjoy mathematics. By asking a few natural questions about Sudoku puzzles, you are led inevitably to major ideas in combinatorics, number theory, and algebra. Sudoku puzzles forced us to confront the changes to mathematical practice wrought by the ready availability of computers, as well as the philosophical issues raised thereby. And we could easily write a second book with all of the material we left out of this one.

Not too shabby for a mere pencil puzzle.

10

Epilogue

You Can Never Have Too Many Puzzles

We leave you with a tour of Sudoku variations. There is a mighty treatise to be written on the mathematical aspects of what follows, but we shall leave that for later. For now, just get out your pencils and play.

10.1 EXTRA REGIONS

We open with a pair of puzzles involving additional Sudoku regions, reminiscent of the Four Square puzzle from Chapter 1.

Puzzle 77: Staples.

Fill in the grid so that every row, column, block, and staple-shaped region contains 1–9 exactly once.

		7					2	
		3				7		4
9	2		4			3	6	
		1	2					
					7	2		
	3	5			1		8	6
8		6				4		
	9					5		

Puzzle 78: Pyramids.

Fill in the grid so that every row, column, block, and pyramid region contains 1–9 exactly once.

9			2					
	4			9				
		7			8			2
7			1				5	
		1		2		7		
	2				4			9
4			5			1		
				3			4	
					1			3

Of course, there is no rule requiring our extra regions to be contiguous. The familiar Sudoku X is an example with linear regions, and the next two moonshine-themed puzzles further develop this idea.

Puzzle 79: Lightning.

Fill in the grid so that every row, column, block, and "lightning bolt" contains 1–9 exactly once.

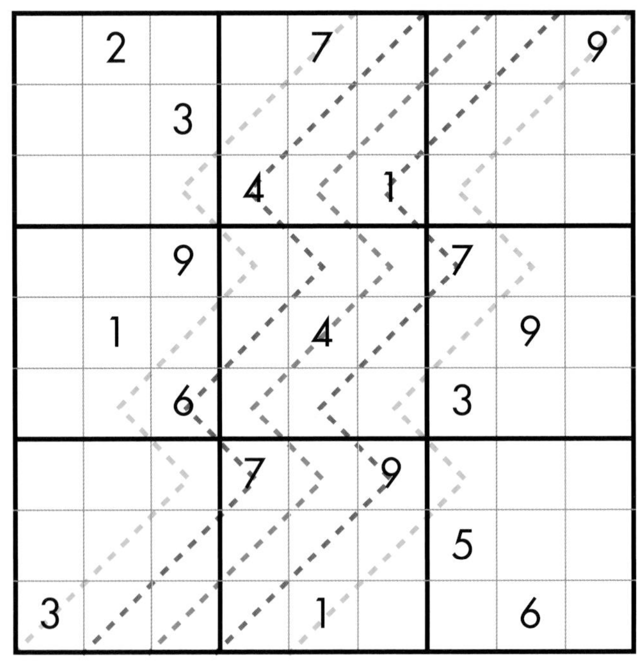

Puzzle 80: XXX.

Fill in the gird so that every row, column, and block contains 1–9 exactly once, and each marked diagonal line contains no repeated entries.

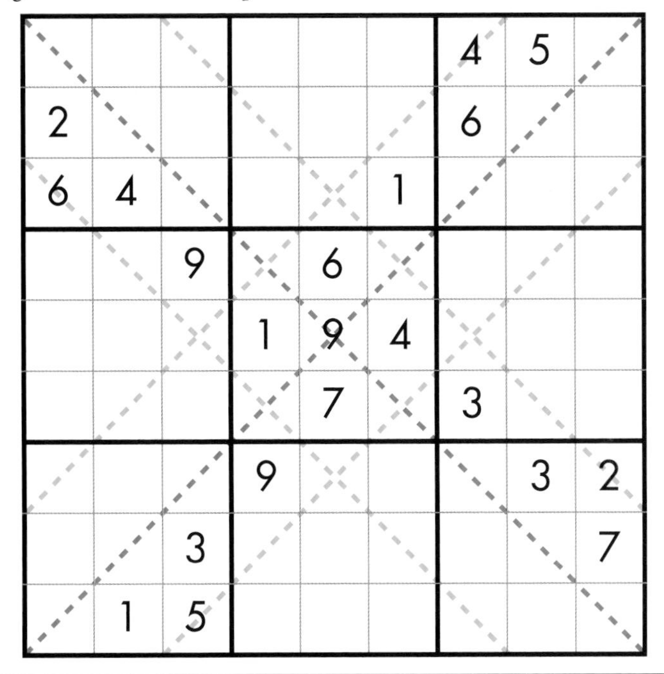

Notice that the Sudoku XXX puzzle above contained *partial regions,* with fewer than nine cells. These regions are too small to contain all nine of the numbers from 1 to 9, but we can require that none of the numbers are repeated within the region. Here are two more partial-region, sock-themed, puzzles:

Puzzle 81: Argyle.

Fill in the grid so that every row, column, and block contains 1–9 exactly once, and each marked diagonal line contains no repeated entries.

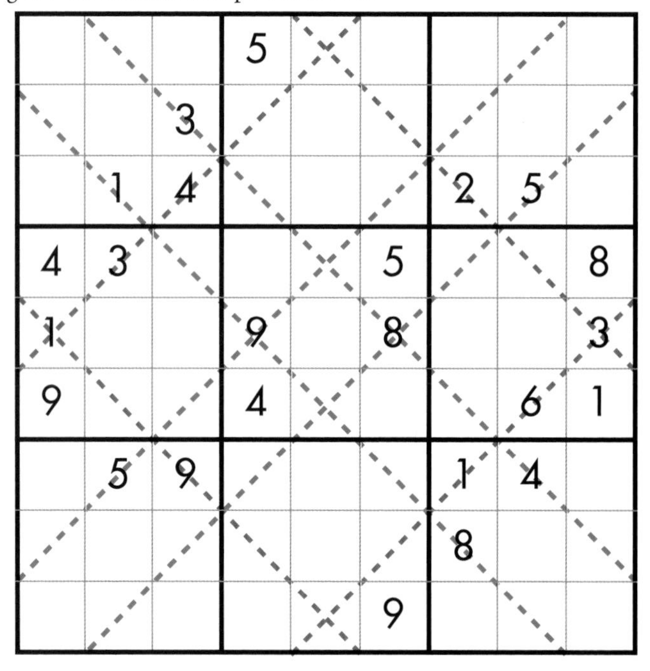

Puzzle 82: Holes.

Fill in the grid so that every row, column, and block contains 1–9 exactly once, and each hole-shaped region contains no repeated entries.

				1		4		
3	1	9		4		2		
7		5						
8	2	3				1	7	6
						3		2
		4		5		6	2	3
		6		2				

In Chapter 6 we saw a puzzle called Jigsaw Plus in which nine additional jigsaw-puzzle-piece regions complemented the standard Sudoku regions. We could also *replace* the blocks with the jigsaw regions, as in the following two puzzles:

Puzzles 83 and 84: Jigsaw Sudoku.

Fill in each grid so that every row, column, and jigsaw-shaped region contains 1–9 exactly once.

						3	4	
	5						8	3
4	1	3						
		1			3			
		5	6	9	1	4		
			4			5		
						9	3	4
3	2						1	
	4	6						

		5				4	9	3
			4				7	
1				2	6	5		
	3				1			
		3				2		
			3				5	
		2	8	4				1
	2				3			
3	4	9				6		

10.2 ADDING VALUE

Standard Sudoku requires nine distinct symbols; that those symbols are digits is purely a matter of convention. As we have noted previously, they could as easily be letters of the alphabet or something more exotic still. What are the possibilities for Sudoku variations in which the numerical values actually matter?

One possibility involves placing an arithmetic condition on the cells. For example, we could require that some or all of the 3×3 blocks are *semimagic squares*, meaning that each of the small rows and small columns in those blocks sum to the same number. The "semi" indicates that we are not requiring the diagonals of the blocks to have the same sum. For example, here is a 3×3 semimagic square using the numbers 1–9 exactly once. The rows and columns add to 15:

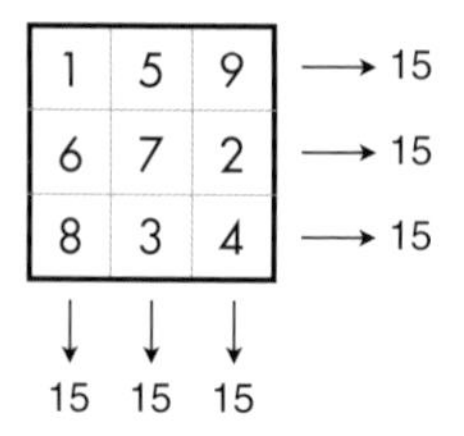

1	5	9	→ 15
6	7	2	→ 15
8	3	4	→ 15
↓ 15	↓ 15	↓ 15	

There are actually very few 3×3 semimagic squares of this sort. The one above is in fact the only one (up to rotation and reflection) whose center cell contains a 7. This means that requiring blocks to be semimagic is in fact a very strong condition.

Solving the following puzzle calls for some preliminary detective work regarding the possibilities for semimagic squares. Here is a hint to get you started: Try to determine the sole possible value for the row sums and column sums of any such square.

Puzzles 85: Three-Magic Sudoku.

Fill in each grid so that every row, column, and block contains 1–9 exactly once. In addition, each of the three shaded blocks must be semimagic squares. That is, their rows and columns add to the same number.

				2				
	2		3		5		8	
				8				
1							5	
8		3				4		6
	9							8
				6				
	4		5		7		6	
				4				

Is it possible to require that *every* block be semimagic? The answer is yes! This condition is so strong that a valid puzzle of this type is possible with just eight starting clues. Our next puzzle provides an example. Here is a hint: Each full row must contain three different small rows that sum to 15. How many ways can you break up the numbers 1–9 into three sets of three numbers so that each set adds to 15? The answer is in the back of the book.

Puzzles 86: All-Magic Sudoku.

Fill in the grid so that every row, column, and block contains 1–9 exactly once. In addition, all nine of the blocks must be semimagic squares whose rows and columns add to the same number. Watch out; this puzzle is challenging and requires its own preliminary detective work about possible semimagic squares.

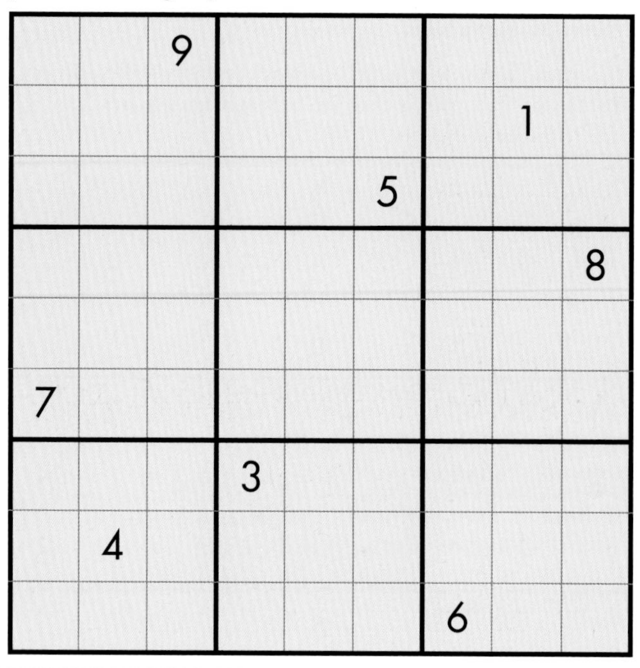

There are other ways of inserting arithmetic into a Sudoku puzzle. We could, for example, just block off various regions and give their required sums. This variation is common and is usually called *Killer Sudoku*:

Puzzles 87: Killer Sudoku.

Fill in the grid so that every row, column, and block contains 1–9 exactly once. In addition, each outlined shape must contain distinct entries whose sum is equal to the number in its upper left corner.

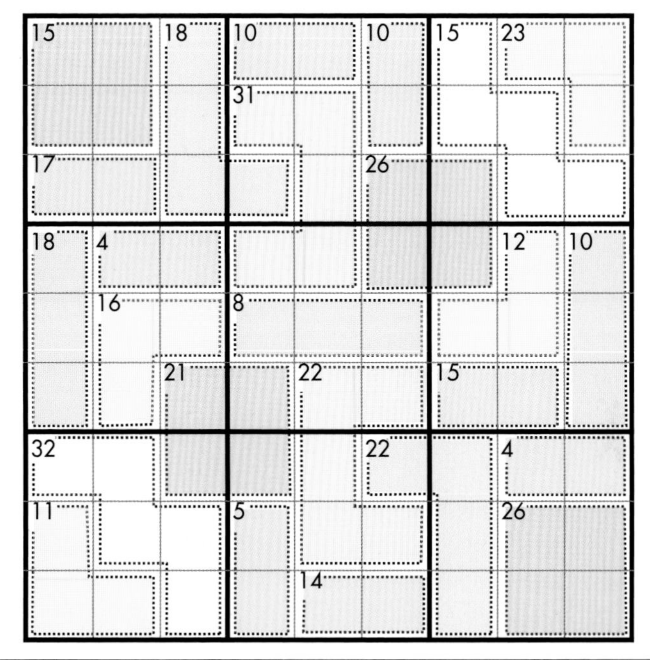

Notice that there were *no* clues in our Killer Sudoku puzzle! The given sums provide so much information that there is only one solution to this puzzle even though no initial clues are provided. If you have not played this sort of puzzle before then getting started can be tricky. If you are desirous of some starting clues, the numbers down the diagonal from the upper left to the lower right are 2, 7, 6, 6, 2, 3, 9, 6, and 5.

Sums are hardly our only option. We could also use products!

Puzzles 88: Product Sudoku.

Fill in the grid so that every row, column, and block contains 1–9 exactly once. In addition, each outlined shape must contain distinct entries whose product is equal to the number in its upper-left corner.

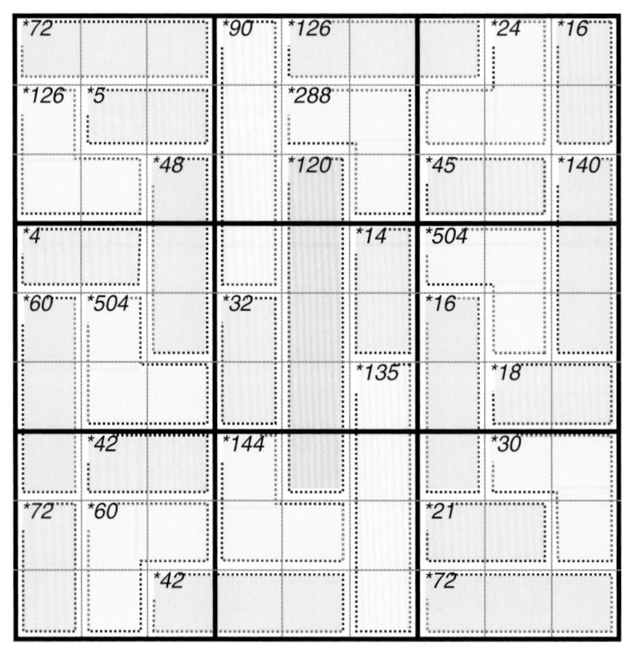

A good starting place for Product Sudoku is to find regions whose products are divisible by 5 or 7. A shape with such a product must contain the digit 5 or 7. Also, note that 1 is potentially a factor of every product. Once again the extra arithmetic conditions are strong enough to force a unique solution despite the absence of initial clues. Product puzzles tend to be easier than those with sums, but if you need a push, here are the entries along the upper-left to lower-right diagonal of the puzzle above: 4, 1, 8, 6, 5, 9, 8, 3, and 9.

10.3 COMPARISON SUDOKU

We could also use simple size comparisons, instead of arithmetic. The carats between the cells in the next two puzzles indicate greater than/less than relationships. (If you have trouble remembering which is the greater than and which the less than symbol, remember that the alligator mouth wants to eat the larger number.)

Puzzle 89: Greater Than Sudoku.

Fill in each grid so that every row, column, and block contains 1–9 exactly once. Also, adjacent cells are greater than or less than their neighbors according to the > or < symbol on the boundary.

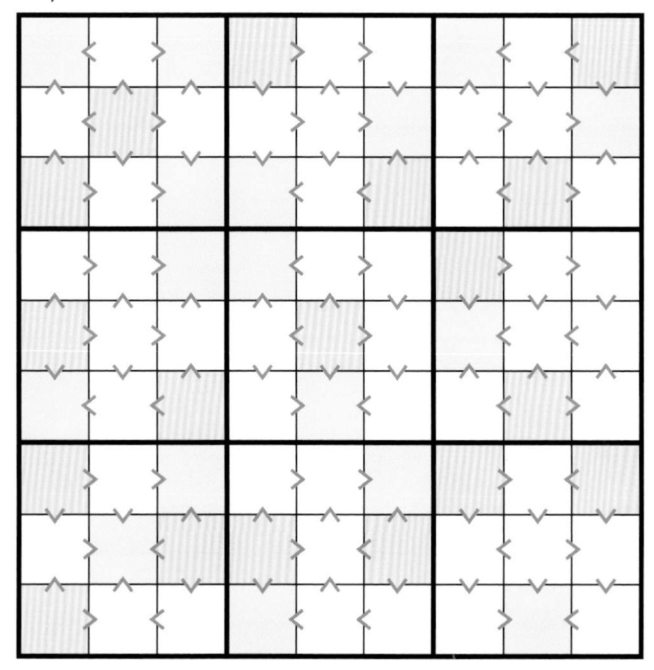

To get started on these puzzles, try to place the 1s and 9s. The pink and blue coloring will help. Within each block, cells are colored pink if they are greater than all of their neighbors, and blue if they are less than all of their neighbors. Since 9s must be greater than all of their neighbors, they can only appear in pink cells. Similarly, 1s can only appear in blue cells. But watch out! Not all pink cells are 9s, and not all blue cells are 1s.

We can also be a little more specific about our "greater than" information, as in the next puzzle:

Puzzle 90: Greater Than Greater.

Fill in each grid so that every row, column, and block contains 1–9 exactly once. Also, adjacent cells are greater than or less than their neighbors according to the > or < symbol on the boundary. The symbols ▷ and ◁ indicate that the cells differ by more than one.

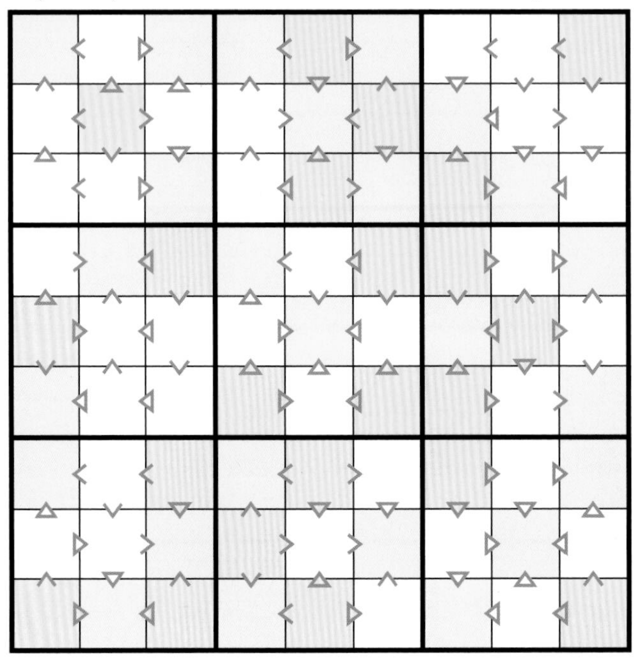

In solving these puzzles, you may have noticed that finding chains of cells, each one greater than the last, is a helpful solving technique. Our next variation focuses on these chains by marking them with "worms."

Puzzles 91 and 92: Worms.

Fill in each grid so that every row, column, and block contains 1–9 exactly once. In addition, each worm must contain entries that increase from tail to head. For blue worms you must figure out yourself which end is the head. Yellow worms contain consecutive numbers, each one exactly one greater than the last.

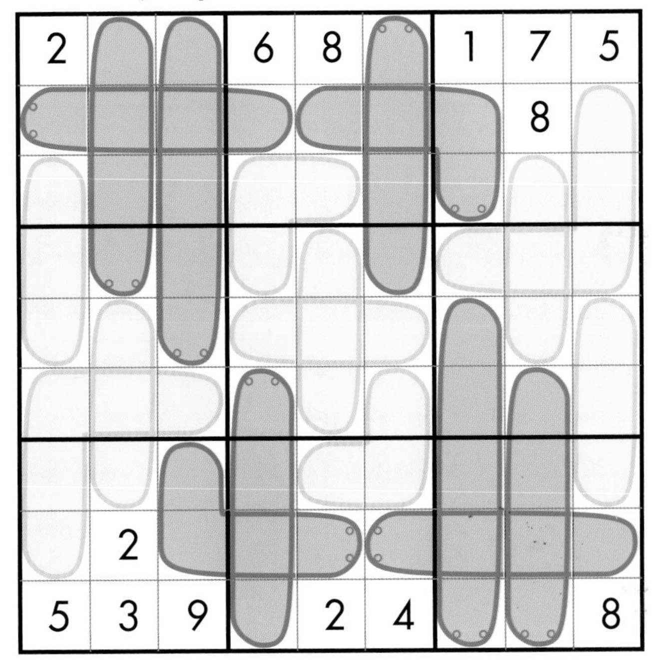

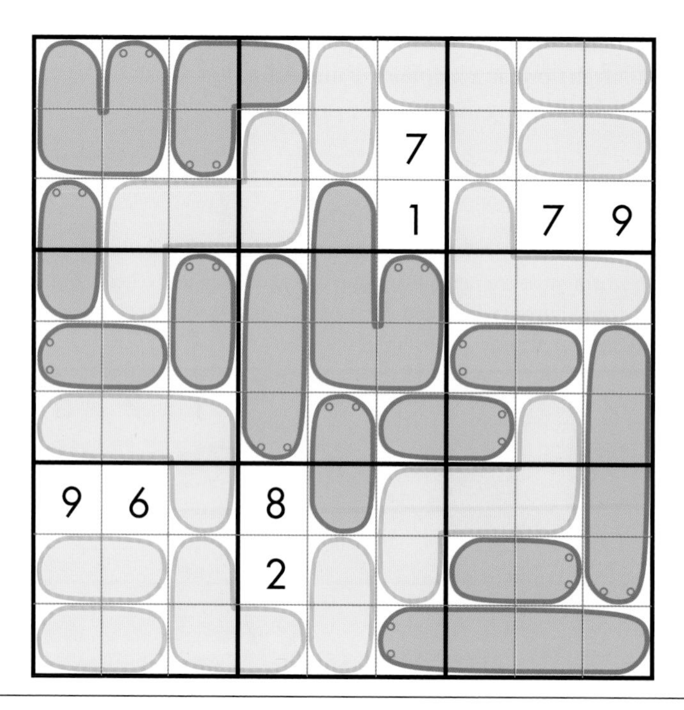
7
1
7
9
9
6
8
2

There are other ways of comparing adjacent cells. We could, for example, indicate divisor relations among neighboring cells.

Puzzle 93: Divided by Sudoku.

Fill in the grid so that every row, column, and block contains 1–9 exactly once. In addition, for neighbors A and B we have $A \subset B$ if A divides B. For example, $2 \subset 4$ and $2 \subset 8$. Note that *all* possible "divided by" relationships are marked in the puzzle. Also, there are a few $<$ and $>$ symbols. As always, these indicate greater than/less than relations.

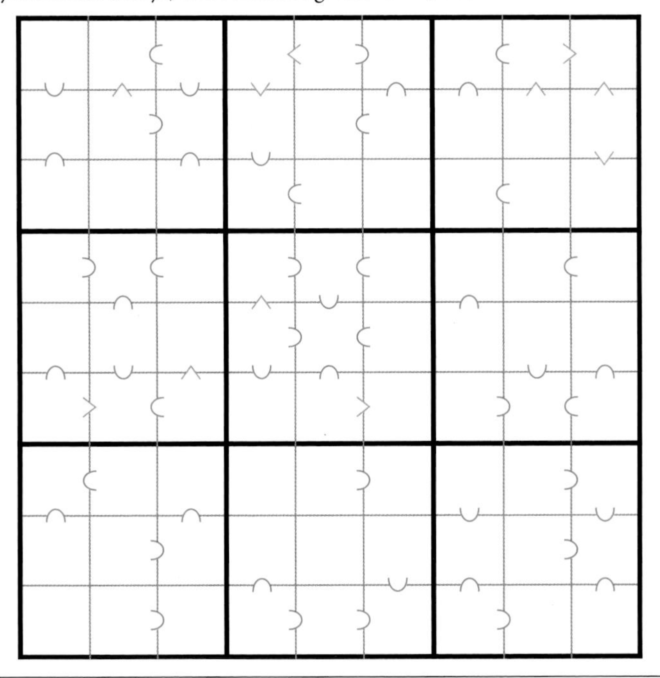

In these puzzles, we must include some greater-than and less-than symbols. Why? Because 5 and 7 have exactly the same divisibility properties: neither is divided by any number other than 1, and neither divides any other number. It would be impossible to distinguish 5s from 7s without the $>$ and $<$ clues. As with the Greater Than Sudoku puzzles, a good place to start is to look for chains of neighbors, each dividing the next. You might also want to ponder the divisibility properties of the numbers 1–9 before starting!

10.4 ... AND BEYOND

To this point, all of our variations added some extra conditions to the usual Sudoku setup. Perhaps we should try rethinking the game altogether.

For example, we do not have to stick to 9×9. We have already explored 4×4 and 6×6, so why not kick it up a notch with a 12×12?

Puzzle 94: Dozendoku.

Fill in the grid so that every row, column, and block contains 1–12 exactly once.

		8	5	11	1			7			2
		6			2		10		5	4	
	11				8	5				6	
				7			12		4		9
		1	4			2					12
		11			5					1	3
7	9					10			6		
6					4			5	2		
11		10		8			9				
	7				10	6				3	
	8	9		5		12			11		
5			11			8	1	4	12		

In Chapter 6, we considered Venn Sudoku, with three overlapping Sudoku puzzles. The shared regions were constrained by conditions on more than one board. Here the same idea is taken to the next level, with five overlapping puzzles in an arrangement known as a *quincunx*. In addition the following puzzle also adds the familiar diagonal "X" condition:

Puzzle 95: Samurai Sudoku X.

Fill in each of the five grids so that every row, column, block, and main diagonal contains 1–9 exactly once.

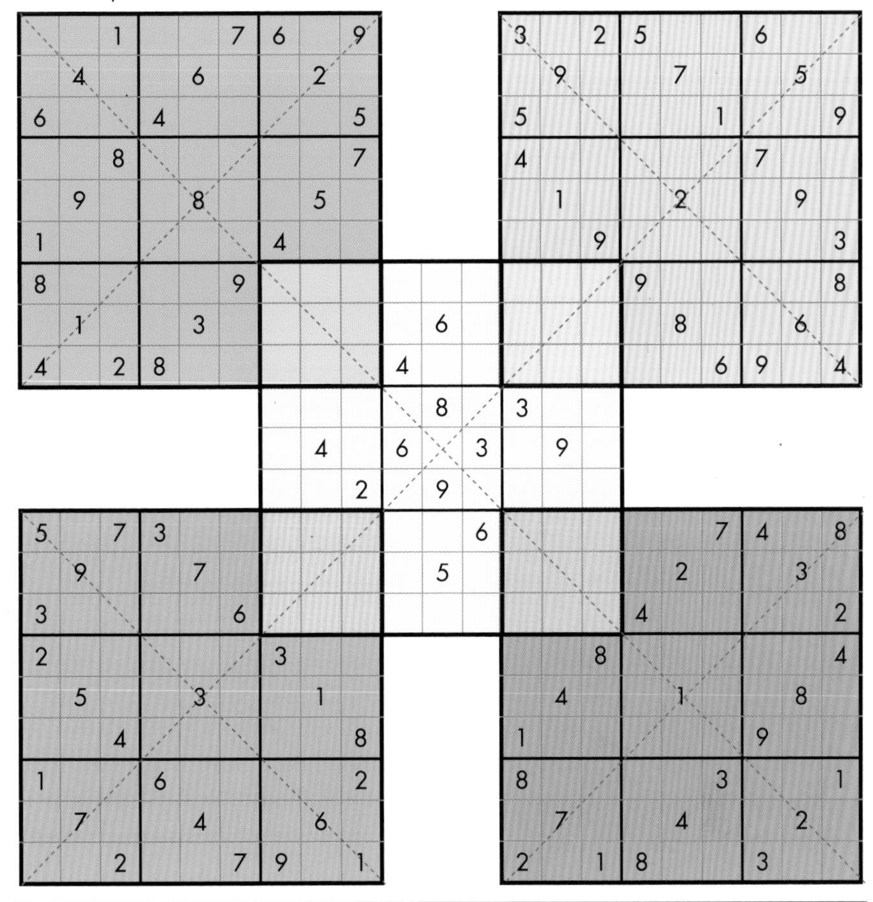

Finally, instead of overlapping run-of-the-mill Sudoku squares, we can change the rules and regions themselves. In Double Trouble Sudoku from Chapter 1 we saw that we could have some values repeated in each row, column, and block. The concept of repeated entries opens up the field quite a bit. The following 8×8 variation has blocks that are jigsaw-type regions containing 1–4, while the rows and columns are *double* regions in which each number appears *twice*.

Puzzle 96: Tetrominoku.

Fill in the grid so that each tetromino shape contains 1–4 exactly once, and each row and column contains 1–4 exactly twice.

1	1				2		2
4	2			1			
		1	2			2	1
			2				
				4			
3	4			2	2		
			1			2	4
4		2				4	3

We called the puzzle above "Tetrominoku" because our jigsaw-shaped regions are known among mathematicians as *tetrominos*. They are a special case of *polyominos*, which are connected figures built by adjoining 1×1 squares along their edges. A tetromino is built from four such squares. There are only five tetromino shapes, up to reflection and rotation. Four of them are found in our puzzle; what is the fifth?

Here is a puzzle built from *pentominos*, which use five squares:

Puzzle 97: Pentominoku.

Fill in the grid so that each pentomino shape contains 1–5 exactly once, and each row and column contains 1–5 exactly twice.

2			4			3			4
	4							5	
		3					2		
2			1			1			3
				4	3				
				3	2				
4			1			2			5
		4					3		
	2							5	
1			5			4			1

Get the idea? We are constrained only by our imaginations, and we could easily have filled this book with nothing but further Sudoku novelties. Even better is that each new variation brings with it new mathematical questions. Should we choose to write a sequel to this volume, we certainly will not be hurting for material.

Joseph de Maistre, a French philosopher active in the late eighteenth century, once remarked, "It is one of man's curious idiosyncrasies to create difficulties for the pleasure of resolving them." The world of professional mathematics owes its existence to the truth of that maxim. We are curious for a living. Asked to find an arrangement of eighty-one digits satisfying some set of rules or whatnot, it does not even occur to us to ask why we would do such a thing. We have been challenged; that is enough. And after finding the required placement, we instinctively ask how many other arrangements there could be, or what happens when we change the rules, or countless other questions in a process, we are happy to report, that simply never ends.

SOLUTIONS TO PUZZLES

Solutions to puzzles that are not in the body of the text can be found here.

1. Sudoku Warm-Up

7	2	3	8	6	5	9	4	1
1	5	9	7	4	2	6	8	3
8	6	4	1	3	9	2	5	7
5	9	8	3	7	4	1	6	2
2	7	1	9	8	6	5	3	4
4	3	6	5	2	1	7	9	8
3	4	5	6	1	7	8	2	9
6	1	2	4	9	8	3	7	5
9	8	7	2	5	3	4	1	6

2. Eulerian Circuits

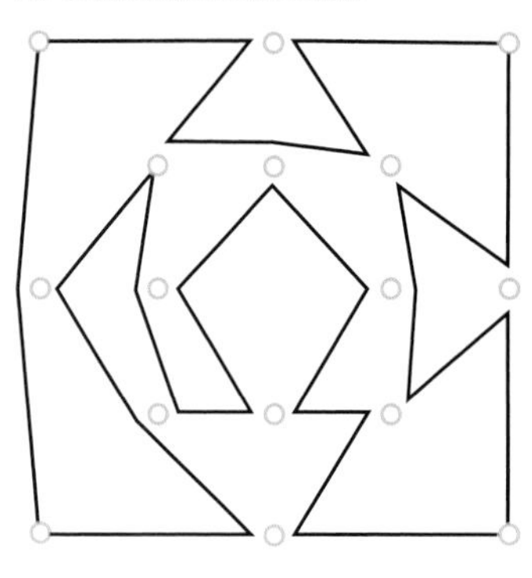

3. Hamiltonian Circuits

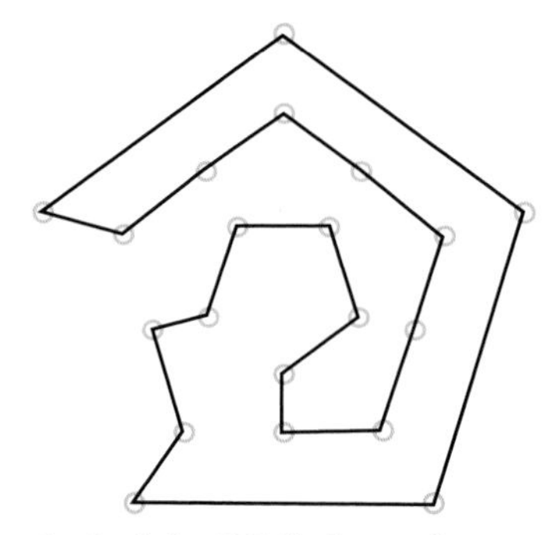

4. Sudoku Walkthrough

7	6	3	5	1	4	2	9	8
5	4	8	9	3	2	7	1	6
2	1	9	7	8	6	5	4	3
8	5	6	1	4	7	3	2	9
3	7	4	2	6	9	8	5	1
9	2	1	8	5	3	6	7	4
6	8	7	4	2	1	9	3	5
1	3	2	6	9	5	4	8	7
4	9	5	3	7	8	1	6	2

5. Harder Sudoku

6	7	8	1	9	5	2	3	4
9	4	5	3	2	7	8	6	1
3	1	2	8	4	6	5	7	9
8	9	6	5	7	3	1	4	2
5	2	7	4	8	1	3	9	6
1	3	4	9	6	2	7	5	8
7	6	9	2	5	8	4	1	3
2	5	1	6	3	4	9	8	7
4	8	3	7	1	9	6	2	5

8. Three Star Sudoku

3	★	6	★	2	1	★	5	4
★	1	2	5	★	4	6	★	3
5	★	4	3	6	★	★	2	1
★	5	★	2	4	★	1	3	6
1	6	★	★	3	★	2	4	5
4	2	3	1	5	6	★	★	★
6	3	★	★	1	5	4	★	2
★	4	5	6	★	2	3	1	★
2	★	1	4	★	3	5	6	★

6. Sudoku X

9	2	7	4	3	6	8	1	5
3	8	6	1	9	5	4	7	2
1	4	5	2	7	8	6	9	3
2	5	3	7	8	4	9	6	1
7	6	8	5	1	9	2	3	4
4	1	9	3	6	2	7	5	8
5	7	2	6	4	1	3	8	9
6	9	1	8	2	3	5	4	7
8	3	4	9	5	7	1	2	6

9. Double Trouble Sudoku

5	3	6	6	4	1	2	4	2
4	4	1	2	2	5	6	6	3
6	2	2	4	3	6	1	5	4
2	5	4	6	6	2	3	1	4
3	6	4	5	1	2	6	4	2
1	2	6	3	4	4	5	2	6
6	1	3	4	6	4	2	2	5
2	4	5	1	2	3	4	6	6
4	6	2	2	5	6	4	3	1

7. Four Square Sudoku

5	4	2	1	8	7	9	6	3
6	7	8	2	3	9	1	5	4
9	3	1	4	5	6	2	7	8
7	6	9	5	1	3	4	8	2
4	8	3	6	9	2	5	1	7
1	2	5	7	4	8	3	9	6
8	9	6	3	2	1	7	4	5
3	1	4	8	7	5	6	2	9
2	5	7	9	6	4	8	3	1

13. Latin-Doku

4	7	6	5	1	2	3	8	9
9	5	8	7	2	3	1	4	6
5	9	2	8	3	4	6	7	1
7	6	1	4	9	8	5	3	2
6	8	3	1	4	5	9	2	7
8	2	9	6	5	7	4	1	3
3	1	7	9	8	6	2	5	4
2	4	5	3	6	1	7	9	8
1	3	4	2	7	9	8	6	5

14 and 15. Bomb Sudoku

3	5	9	4	1	8	2	6	7
7	1	2	6	3	4	5	9	8
2	4	3	5	7	9	8	1	6
5	9	7	8	2	1	6	4	3
1	6	4	3	5	7	9	8	2
4	2	8	1	9	6	3	7	5
8	3	6	7	4	5	1	2	9
6	7	5	9	8	2	4	3	1
9	8	1	2	6	3	7	5	4

5	2	3	4	1	8	9	7	6
3	1	7	2	5	4	6	8	9
8	9	4	6	3	7	2	5	1
2	7	1	8	9	5	4	6	3
9	5	6	3	4	1	8	2	7
7	8	9	1	2	6	5	3	4
4	6	2	5	7	3	1	9	8
1	3	8	9	6	2	7	4	5
6	4	5	7	8	9	3	1	2

18. Greco-Latin-Doku

A 1	C 4	B 3	E 5	D 2
D 4	B 2	A 5	C 1	E 3
C 3	D 5	E 4	A 2	B 1
E 2	A 3	D 1	B 4	C 5
B 5	E 1	C 2	D 3	A 4

19. Crossdoku

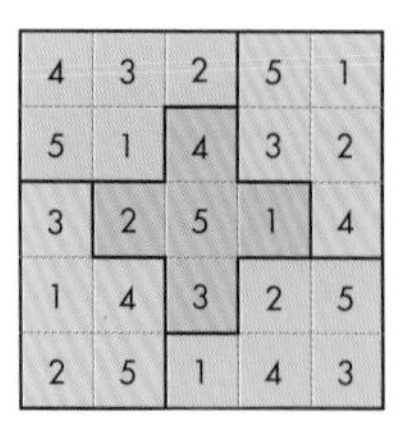

4	3	2	5	1
5	1	4	3	2
3	2	5	1	4
1	4	3	2	5
2	5	1	4	3

3	1	2	5	4
5	4	3	1	2
1	2	5	4	3
4	3	1	2	5
2	5	4	3	1

20. Related Crossdoku Cells

The upper-right blue region needs an *X*. The *X* in the central cross precludes placing an *X* in the second row of the blue region. If we place an *X* in the first row of the blue region, then there is no place to put an *X* in the upper-left green region. The only remaining possibility is to place an *X* in the cell with the question mark.

21. Greco-Latin Mini-Sudoku

B 3	A 2	C 1	D 4
C 4	D 1	B 2	A 3
A 1	B 4	D 3	C 2
D 2	C 3	A 4	B 1

A 4	C 3	B 1	D 2
B 2	D 1	A 3	C 4
C 1	A 2	D 4	B 3
D 3	B 4	C 2	A 1

22. Greco-Latin Sudoku

E 7	H 2	A 5	I 8	B 3	F 9	G 1	C 6	D 4
G 6	C 4	D 1	E 2	H 5	A 7	I 3	B 9	F 8
I 9	B 8	F 3	G 4	C 1	D 6	E 5	H 7	A 2
C 8	D 9	E 4	A 1	I 7	B 5	F 2	G 3	H 6
A 3	G 5	I 6	H 9	F 4	C 2	D 7	E 8	B 1
H 1	F 7	B 2	D 3	E 6	G 8	C 9	A 4	I 5
F 5	A 6	G 9	B 7	D 8	E 1	H 4	I 2	C 3
D 2	E 3	H 8	C 5	A 9	I 4	B 6	F 1	G 7
B 4	I 1	C 7	F 6	G 2	H 3	A 8	D 5	E 9

24. Checkerboard Squares

There are twenty-five 1×1 squares, sixteen 2×2 squares, nine 3×3 squares, four 4×4 squares, and one 5×5 square, for a total of 55 squares.

25. Tournament Matches

In the first tournament, sixty-three matches must be played. In the second tournament, seventy-nine matches must be played.

26. Shidoku

4	2	1	3
3	1	2	4
2	3	4	1
1	4	3	2

3	2	1	4
4	1	2	3
2	3	4	1
1	4	3	2

4	3	1	2
2	1	4	3
3	4	2	1
1	2	3	4

27. Ordered Shidoku Boards

1	2	3	4
3	4	1	2
2	1	4	3
4	3	2	1

1	2	3	4
3	4	1	2
2	3	4	1
4	1	2	3

1	2	3	4
3	4	2	1
2	1	4	3
4	3	1	2

29. Completing Shidoku

There are four ways to complete the first configuration, and two ways to complete the second configuration. These six ways are:

1	2	3	4
3	4	1	2
2	1	4	3
4	3	2	1

1	2	3	4
3	4	1	2
2	3	4	1
4	1	2	3

1	2	3	4
3	4	1	2
4	1	2	3
2	3	4	1

1	2	3	4
3	4	1	2
4	3	2	1
2	1	4	3

1	2	3	4
3	4	2	1
2	1	4	3
4	3	1	2

1	2	3	4
3	4	2	1
4	3	1	2
2	1	4	3

30 and 31. Sudoku Reward

2	7	6	9	3	8	1	4	5
3	8	4	1	5	2	6	9	7
5	9	1	4	6	7	2	8	3
1	4	5	3	8	9	7	2	6
9	6	2	5	7	4	8	3	1
7	3	8	2	1	6	4	5	9
6	1	3	8	2	5	9	7	4
8	5	9	7	4	1	3	6	2
4	2	7	6	9	3	5	1	8

2	1	9	3	7	4	6	5	8
8	4	7	9	6	5	1	2	3
5	3	6	8	1	2	9	7	4
4	6	5	1	3	7	8	9	2
9	7	3	6	2	8	5	4	1
1	8	2	5	4	9	7	3	6
3	5	1	2	9	6	4	8	7
7	2	8	4	5	1	3	6	9
6	9	4	7	8	3	2	1	5

33. Shidoku Equivalence

The first square is type 1; relabel by the rule $1 \rightarrow 2 \rightarrow 4 \rightarrow 3 \rightarrow 1$, and then swap the last two columns.

The second square is type 2; flip over the diagonal from the upper left to the lower right, relabel the numbers by $3 \leftrightarrow 4$, and swap the last two columns.

The third square is also type 2; relabel by $2 \rightarrow 4$, $3 \rightarrow 2$, and $4 \rightarrow 3$ and then rotate ninety degrees counterclockwise.

34. Composing Symmetries

Transforming by the reflection a and then the reflection b is the same as the rotation R by 240 degrees. Transforming in the other order with reflection b followed by reflection a is the same as the rotation r by 120 degrees.

36. Burnside's Shidoku Transpose

1	2	3	4
3	4	1	2
2	1	4	3
4	3	2	1

1	2	4	3
3	4	2	1
4	3	1	2
2	1	3	4

37 and 38. Sudoku Clones

8	6	9	5	3	4	2	1	7
1	3	7	6	9	2	8	5	4
5	4	2	1	8	7	9	6	3
6	5	4	3	1	8	7	2	9
7	9	3	2	6	5	4	8	1
2	1	8	4	7	9	6	3	5
4	2	6	9	5	3	1	7	8
9	8	5	7	2	1	3	4	6
3	7	1	8	4	6	5	9	2

6	1	4	9	7	3	2	5	8
9	8	7	5	1	2	3	6	4
3	5	2	8	6	4	7	9	1
1	2	8	6	5	9	4	3	7
5	9	3	1	4	7	6	8	2
4	7	6	3	2	8	5	1	9
8	4	9	7	3	6	1	2	5
2	3	5	4	9	1	8	7	6
7	6	1	2	8	5	9	4	3

As a matter of principle, we cannot reveal the solutions to Bonus Round problems! Ha ha!

39 and 40. Roku-Doku

1	6	3	2	5	4
2	5	4	6	3	1
3	1	6	4	2	5
5	4	2	1	6	3
4	2	5	3	1	6
6	3	1	5	4	2

4	6	2	3	5	1
5	3	1	6	4	2
3	1	4	5	2	6
6	2	5	1	3	4
2	5	6	4	1	3
1	4	3	2	6	5

43. The Eighteen-Clue Needle

7	8	1	4	6	5	3	9	2
5	4	2	3	1	9	8	6	7
6	3	9	2	8	7	1	5	4
4	6	3	5	9	2	7	1	8
9	2	8	1	7	3	5	4	6
1	5	7	6	4	8	2	3	9
8	1	6	7	5	4	9	2	3
2	7	4	9	3	1	6	8	5
3	9	5	8	2	6	4	7	1

44 and 45. Two More Needles

5	7	4	3	9	8	6	1	2
1	3	2	5	4	6	9	7	8
8	9	6	2	1	7	4	5	3
2	8	9	1	3	4	7	6	5
3	4	5	6	7	2	1	8	9
6	1	7	9	8	5	2	3	4
4	5	8	7	6	9	3	2	1
7	2	1	4	5	3	8	9	6
9	6	3	8	2	1	5	4	7

1	9	3	7	6	5	2	4	8
8	6	5	2	3	4	1	9	7
2	7	4	8	1	9	5	3	6
6	2	9	3	4	8	7	1	5
4	8	1	5	7	2	3	6	9
5	3	7	6	9	1	8	2	4
3	1	8	9	5	6	4	7	2
7	5	6	4	2	3	9	8	1
9	4	2	1	8	7	6	5	3

46. Eighteen-Clue Pi

7	2	9	1	4	6	3	5	8
3	5	8	7	2	9	1	4	6
4	1	6	5	3	8	7	9	2
8	9	7	4	1	2	5	6	3
2	4	3	8	6	5	9	7	1
5	6	1	9	7	3	8	2	4
9	3	4	2	5	1	6	8	7
1	8	2	6	9	7	4	3	5
6	7	5	3	8	4	2	1	9

47 and 48. Easy Twenty and Hard Twenty-Eight

3	6	7	5	1	2	8	9	4
8	9	2	4	7	3	5	1	6
4	5	1	6	9	8	3	2	7
7	1	9	8	2	5	4	6	3
2	4	6	1	3	7	9	5	8
5	3	8	9	6	4	2	7	1
6	8	4	2	5	1	7	3	9
9	2	3	7	8	6	1	4	5
1	7	5	3	4	9	6	8	2

8	4	3	7	5	2	9	6	1
6	2	5	9	1	8	4	3	7
1	9	7	3	4	6	8	5	2
3	6	1	5	8	9	2	7	4
7	5	9	6	2	4	1	8	3
4	8	2	1	7	3	5	9	6
5	3	6	4	9	1	7	2	8
9	1	8	2	6	7	3	4	5
2	7	4	8	3	5	6	1	9

49 and 50. Easy and Hard Twins

8	3	1	7	6	5	4	9	2
4	9	6	1	2	8	3	7	5
7	5	2	4	9	3	1	8	6
3	1	8	9	5	4	6	2	7
9	6	4	3	7	2	8	5	1
2	7	5	8	1	6	9	3	4
1	4	7	2	8	9	5	6	3
6	8	3	5	4	7	2	1	9
5	2	9	6	3	1	7	4	8

9	8	4	3	6	2	5	1	7
2	1	6	5	7	4	3	8	9
7	5	3	9	1	8	4	2	6
1	2	8	7	3	5	9	6	4
6	4	7	8	9	1	2	3	5
5	3	9	4	2	6	1	7	8
8	9	2	1	4	7	6	5	3
4	7	1	6	5	3	8	9	2
3	6	5	2	8	9	7	4	1

51. Checkerboard Dominos

No, it is not possible. Each domino must cover two adjacent squares on the checkerboard, and adjacent squares are always different colors: one white and one black. This means that any set of dominos always covers the same number of white squares as black squares.

Since the opposite corners of a checkerboard are always the same color, the remaining squares have an unequal amount of white and black squares and therefore cannot be covered by dominos.

52. Chessboard Knights

The most that can be placed is thirty-two. A knight always moves from a white square to a black square or vice versa. Therefore we can put the thirty-two knights on the white squares and no two will attack each other.

53. Jigsaw Plus

6	3	7	9	5	1	8	2	4
5	1	4	8	3	2	9	4	6
2	9	8	7	4	6	5	1	3
1	4	2	6	7	5	3	8	9
8	6	5	3	1	9	2	4	7
3	7	9	2	8	4	1	6	5
9	2	3	4	6	8	7	5	1
4	8	1	5	9	7	6	3	2
7	5	6	1	2	3	4	9	8

54. Rainbow Wrap

7	2	3	4	1	8	6	5	9
8	9	1	6	3	5	7	4	2
4	6	5	9	7	2	1	8	3
5	7	4	2	8	6	3	9	1
2	1	8	3	5	9	4	6	7
6	3	9	7	4	1	5	2	8
9	4	2	1	6	7	8	3	5
1	8	6	5	2	3	9	7	4
3	5	7	8	9	4	2	1	6

55. Position Sudoku

2	5	9	7	3	4	1	6	8
7	3	6	1	8	9	4	5	2
4	8	1	2	5	6	3	9	7
3	2	5	4	9	7	6	8	1
9	7	8	6	1	3	2	4	5
1	6	4	8	2	5	7	3	9
8	1	3	5	4	2	9	7	6
5	9	7	3	6	1	8	2	4
6	4	2	9	7	8	5	1	3

56. Venn Sudoku

2	9	5	1	4	7	6	3	8						
8	3	1	2	5	6	4	9	7						
4	6	7	8	9	3	1	5	2						
1	4	3	9	2	5	7	8	6	2	9	3	5	4	1
9	5	2	7	6	8	3	1	4	5	6	7	2	9	8
7	8	6	3	1	4	5	2	9	4	8	1	6	7	3
3	2	8	4	7	1	9	6	5	3	2	8	4	1	7
6	7	9	5	3	2	8	4	1	9	7	6	3	2	5
5	1	4	6	8	9	2	7	3	1	4	5	8	6	9
			1	6	3	4	9	7	8	5	2	1	3	6
			9	2	5	6	3	8	7	1	4	9	5	2
			7	4	8	1	5	2	6	3	9	7	8	4
			2	1	4	7	8	9	5	6	3			
			8	5	6	3	1	4	2	9	7			
			3	9	7	5	2	6	4	8	1			

57. Jigsaw Pi Sudoku

3	2	5	1	5	4	6	3	1	8	9	5
4	1	5	2	3	8	5	9	5	1	3	6
6	1	4	5	9	3	5	8	3	1	2	5
5	3	3	1	8	5	9	2	5	6	4	1
8	9	2	6	5	1	1	5	4	3	3	5
5	8	1	5	2	9	4	3	3	5	6	1
1	5	3	8	1	6	2	4	9	5	5	3
9	4	5	3	5	1	5	6	8	2	1	3
2	3	6	5	1	5	3	1	5	4	8	9
3	6	8	9	4	5	1	5	1	3	5	2
1	5	1	3	6	3	8	5	2	9	5	4
5	5	9	4	3	2	3	1	6	5	1	8

58. Three-Coloring Graphs

The Petersen and dodecahedron graphs are properly three-colorable. Examples of how to color them are below. The Grötzsch graph is famously not three-colorable.

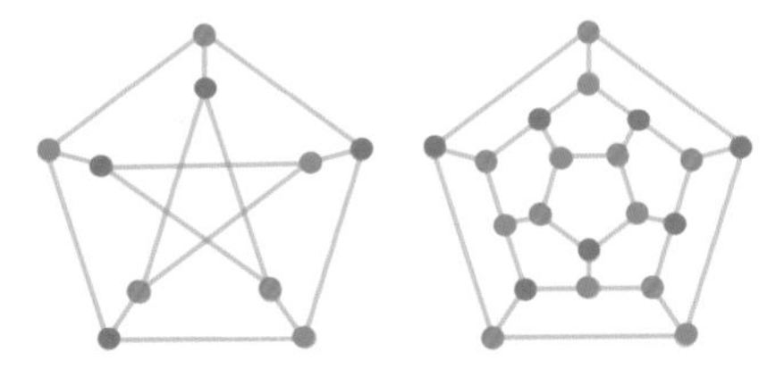

59. Edge-Coloring Graphs

The only difficult part is figuring out which one of them can be edge-colored with just three colors; that is the dodecahedron graph, shown below. (This edge coloring is related to the Hamiltonian circuit found in puzzle 3; can you see how?)

It is not difficult to properly edge-color the Petersen graph with four colors, and the Grötzsch graph with five. What is less obvious is why these are the fewest possible colors that work!

60. Triangles in Complete Graphs

It is easy to color the edges of K_5 red or blue without making any monochromatic triangles; just color the inner star red and the outer pentagon blue.

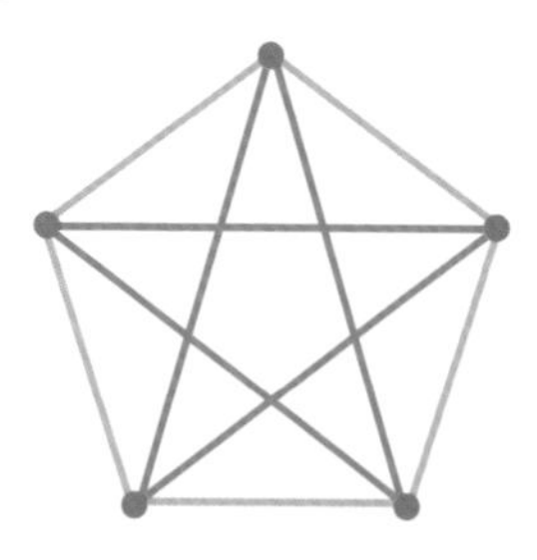

To see that K_6 must contain at least one monochromatic triangle, start by picking any vertex. That vertex must be connected to all five other vertices. Since we have five edges and two colors, we see that at least three of those edges must be the same color, let us say red. Those three red edges connect our original vertex to three other vertices. If the three edges joining those three other vertices are all blue, then we have a monochromatic triangle. Otherwise, there is at least one red edge among those three other vertices. But then the endpoints of that edge, together with our original vertex, form an all-red triangle.

61. Four-Coloring America

It is not difficult to color the states with just four colors. One method is to use three colors only for as long as possible, only using the fourth when you get into a bind. For this particular map it is not difficult.

But could you ever do this without needing the fourth color? In this case, no. Look at Nevada. It is surrounded by five states and must be a different color from all five. Suppose Nevada is blue, and we color Oregon red. Moving clockwise we could alternate colors with Idaho in green, Utah in red, and Arizona in green. Now what to color California? Here is a place where we must have a fourth color.

62. Four-Color Wheels

The pattern is that if the center county is surrounded by an even number of counties, then three colors suffice, by alternating colors around the wheel. If a county is surrounded by an odd number of counties (greater than or equal to three), then four colors are needed, since alternating colors will no longer work. Note that this is exactly what happened with Nevada in the previous puzzle.

65. Bad News Sudoku

4	5	9	2	7	1	9	4	4
4	9	7	5	4	4	2	1	9
1	2	4	9	9	4	4	5	7
5	4	9	7	1	4	4	9	2
7	4	1	9	2	9	5	4	4
2	9	4	4	4	5	7	9	1
4	1	4	4	9	2	9	7	5
9	4	5	4	4	7	1	2	9
9	7	2	1	5	9	4	4	4

66. Shift Sudoku

5	-2	4	7	2	-1	3	6	1
3	7	2	4	6	1	5	-2	-1
-1	6	1	3	5	-2	7	4	2
2	-1	7	6	4	3	1	5	-2
6	1	-2	5	-1	7	2	3	4
4	3	5	1	-2	2	-1	7	6
-2	4	3	-1	1	5	6	2	7
7	2	-1	-2	3	6	4	1	5
1	5	6	2	7	4	-2	-1	3

67. Complex Shidoku

-1	1	$-i$	i
$-i$	i	-1	1
i	-1	1	$-i$
1	$-i$	i	-1

i	1	-1	$-i$
-1	$-i$	1	i
1	i	$-i$	-1
$-i$	-1	i	1

72. Off-Diagonal Sudoku

7	2	3	1	5	4	9	6	8
9	6	8	3	2	7	5	1	4
5	1	4	9	6	8	7	3	2
1	4	6	2	9	3	8	7	5
3	5	9	7	8	6	2	4	1
2	8	7	4	1	5	3	9	6
8	7	2	6	4	9	1	5	3
4	9	5	8	3	1	6	2	7
6	3	1	5	7	2	4	8	9

73. Empty Space Sudoku

8	6	1	5	3	4	9	2	7
7	9	3	8	1	2	5	6	4
4	5	2	7	6	9	3	8	1
6	3	5	4	2	7	1	9	8
9	1	8	6	5	3	7	4	2
2	4	7	1	9	8	6	3	5
3	7	6	2	8	1	4	5	9
1	8	9	3	4	5	2	7	6
5	2	4	9	7	6	8	1	3

74. Avoidance Sudoku

9	5	6	4	7	8	3	1	2
2	1	8	9	6	3	4	5	7
3	7	4	5	1	2	6	8	9
4	9	5	3	2	6	8	7	1
1	3	7	8	9	5	2	6	4
8	6	2	7	4	1	9	3	5
5	2	9	6	3	7	1	4	8
7	4	3	1	8	9	5	2	6
6	8	1	2	5	4	7	9	3

75. 17-Clue Sudoku

9	3	7	6	4	5	8	2	1
8	5	2	9	1	3	4	7	6
6	1	4	2	8	7	3	5	9
7	6	3	8	2	9	1	4	5
2	4	9	5	3	1	6	8	7
1	8	5	4	7	6	9	3	2
4	9	6	3	5	2	7	1	8
3	2	1	7	9	8	5	6	4
5	7	8	1	6	4	2	9	3

76. Twelve-Clue Sudoku X

7	9	4	5	2	3	8	1	6
8	5	6	7	1	4	2	9	3
2	3	1	8	6	9	4	7	5
9	6	3	4	8	1	5	2	7
1	2	5	9	3	7	6	4	8
4	8	7	2	5	6	1	3	9
3	4	8	6	7	2	9	5	1
6	7	2	1	9	5	3	8	4
5	1	9	3	4	8	7	6	2

77. Staples

1	4	7	6	9	3	8	2	5
5	6	3	8	1	2	7	9	4
9	2	8	4	7	5	3	6	1
3	5	1	2	8	4	6	7	9
4	7	2	9	3	6	1	5	8
6	8	9	1	5	7	2	4	3
2	3	5	7	4	1	9	8	6
8	1	6	5	2	9	4	3	7
7	9	4	3	6	8	5	1	2

78. Pyramids

9	3	5	2	1	7	4	6	8
2	4	8	6	9	5	3	7	1
6	1	7	3	4	8	5	9	2
7	9	4	1	8	3	2	5	6
8	5	1	9	2	6	7	3	4
3	2	6	7	5	4	8	1	9
4	8	3	5	6	9	1	2	7
1	7	9	8	3	2	6	4	5
5	6	2	4	7	1	9	8	3

79. Lightning

6	2	4	5	7	8	1	3	9
1	8	3	6	9	2	4	5	7
9	7	5	4	3	1	6	8	2
8	3	9	1	2	6	7	4	5
7	1	2	3	4	5	8	9	6
5	4	6	9	8	7	3	2	1
4	6	8	7	5	9	2	1	3
2	9	1	8	6	3	5	7	4
3	5	7	2	1	4	9	6	8

80. XXX

3	9	8	7	2	6	4	5	1
2	5	1	4	8	9	6	7	3
6	4	7	3	5	1	8	2	9
5	7	9	8	6	3	2	1	4
8	3	2	1	9	4	7	6	5
1	6	4	5	7	2	3	9	8
7	8	6	9	4	5	1	3	2
9	2	3	6	1	8	5	4	7
4	1	5	2	3	7	9	8	6

81. Argyle

2	6	7	5	9	1	3	8	4
5	9	3	8	2	4	7	1	6
8	1	4	7	3	6	2	5	9
4	3	6	2	1	5	9	7	8
1	7	5	9	6	8	4	2	3
9	8	2	4	7	3	5	6	1
6	5	9	3	8	2	1	4	7
3	4	1	6	5	7	8	9	2
7	2	8	1	4	9	6	3	5

82. Holes

6	5	8	3	1	2	4	9	7
2	4	7	9	6	5	8	3	1
3	1	9	8	4	7	2	6	5
7	6	5	2	3	1	9	4	8
8	2	3	5	9	4	1	7	6
4	9	1	6	7	8	3	5	2
1	8	4	7	5	9	6	2	3
5	3	2	4	8	6	7	1	9
9	7	6	1	2	3	5	8	4

83 and 84. Jigsaw Sudoku

6	8	9	5	7	2	3	4	1
2	5	7	9	6	4	1	8	3
4	1	3	7	8	5	6	2	9
7	6	1	2	4	3	8	9	5
8	3	5	6	9	1	4	7	2
1	9	2	4	3	8	5	6	7
5	7	8	1	2	6	9	3	4
3	2	4	8	5	9	7	1	6
9	4	6	3	1	7	2	5	8

8	1	5	7	6	2	4	9	3
2	9	6	4	3	5	1	7	8
1	7	8	9	2	6	5	3	4
5	3	4	2	7	1	9	8	6
9	8	3	6	5	4	2	1	7
4	6	1	3	9	8	7	5	2
7	5	2	8	4	9	3	6	1
6	2	7	5	1	3	8	4	9
3	4	9	1	8	4	6	2	5

85. Three-Magic Sudoku

First, the answer to the hint before the puzzle: Since we must use the numbers 1–9 exactly once then the sum of all nine cells in the block must be the sum of 1–9, which is 45. That means that each row and column must add to one-third of that sum, which is 15.

9	1	8	4	2	6	7	3	5
4	2	7	3	9	5	6	8	1
3	5	6	7	8	1	2	4	9
1	6	2	8	3	4	9	5	7
8	7	3	1	5	9	4	2	6
5	9	4	6	7	2	3	1	8
7	3	5	2	6	8	1	9	4
2	4	9	5	1	7	8	6	3
6	8	1	9	4	3	5	7	2

86. All-Magic Sudoku

First, the solution to the hint: $1 + 5 + 9$, $2 + 6 + 7$, and $3 + 4 + 8$ is one way, and the only other way is $1 + 8 + 6$, $2 + 9 + 4$, and $3 + 5 + 7$.

4	2	9	6	8	1	3	5	7
3	7	5	2	4	9	8	1	6
8	6	1	7	3	5	4	9	2
2	9	4	5	7	3	1	6	8
6	1	8	9	2	4	5	7	3
7	5	3	1	6	8	9	2	4
1	8	6	3	5	7	2	4	9
9	4	2	8	1	6	7	3	5
5	3	7	4	9	2	6	8	1

87. Killer Sudoku

2	5	4	7	3	6	1	8	9
1	7	3	9	8	4	5	2	6
8	9	6	5	1	2	7	3	4
4	3	1	6	7	9	8	5	2
9	6	8	1	2	5	3	4	7
5	2	7	8	4	3	6	9	1
6	8	2	4	5	7	9	1	3
7	4	5	3	9	1	2	6	8
3	1	9	2	6	8	4	7	5

88. Product Sudoku

4	2	9	5	7	6	3	1	8
7	1	5	3	9	8	6	4	2
3	6	8	1	2	4	5	9	7
1	4	2	6	3	7	9	8	5
6	9	3	8	5	2	1	7	4
5	8	7	4	1	9	2	6	3
2	7	6	9	4	3	8	5	1
9	5	4	2	8	1	7	3	6
8	3	1	7	6	5	4	2	9

89. Greater Than Sudoku

1	8	5	9	6	3	2	4	7
4	9	6	8	7	2	5	3	1
7	3	2	1	4	5	6	9	8
6	2	1	3	8	7	9	5	4
8	7	4	5	9	6	1	2	3
3	5	9	2	1	4	7	8	6
5	4	3	6	2	1	8	7	9
2	1	8	7	3	9	4	6	5
9	6	7	4	5	8	3	1	2

90. Greater Than Greater

3	7	1	2	6	4	5	8	9
4	9	5	3	1	8	2	7	6
6	8	2	7	9	5	4	1	3
5	2	9	1	3	7	8	6	4
7	3	8	4	2	6	1	9	5
1	4	6	8	5	9	7	3	2
2	6	7	5	8	3	9	4	1
8	5	3	9	4	1	6	2	7
9	1	4	6	7	2	3	5	8

91 and 92. Worms

2	4	3	6	8	9	1	7	5
9	7	5	4	1	3	6	8	2
1	8	6	5	7	2	9	4	3
3	9	7	2	6	1	8	5	4
4	1	8	3	5	7	2	9	6
6	5	2	9	4	8	3	1	7
7	6	1	8	3	5	4	2	9
8	2	4	7	9	6	5	3	1
5	3	9	1	2	4	7	6	8

1	9	7	4	8	3	2	5	6
2	5	8	6	9	7	1	3	4
6	3	4	5	2	1	8	7	9
4	2	9	1	3	8	7	6	5
5	1	3	7	4	6	9	8	2
8	7	6	9	5	2	4	1	3
9	6	5	8	1	4	3	2	7
3	4	1	2	7	5	6	9	8
7	8	2	3	6	9	5	4	1

93. Divided by Sudoku

4	3	9	7	8	2	1	6	5
2	5	1	4	3	6	7	9	8
8	6	7	1	5	9	2	4	3
7	1	5	6	2	8	4	3	9
3	4	6	9	1	5	8	7	2
9	2	8	3	7	4	5	1	6
1	7	2	5	6	3	9	8	4
6	8	4	2	9	7	3	5	1
5	9	3	8	4	1	6	2	7

94. Dozendoku

10	4	8	5	11	1	9	6	7	3	12	2
12	1	6	9	3	2	7	10	8	5	4	11
2	11	7	3	12	8	5	4	10	9	6	1
3	2	5	10	7	11	1	12	6	4	8	9
8	6	1	4	10	9	2	3	11	7	5	12
9	12	11	7	6	5	4	8	2	10	1	3
7	9	4	1	2	12	10	5	3	6	11	8
6	3	12	8	1	4	11	7	5	2	9	10
11	5	10	2	8	6	3	9	12	1	7	4
1	7	2	12	4	10	6	11	9	8	3	5
4	8	9	6	5	3	12	2	1	11	10	7
5	10	3	11	9	7	8	1	4	12	2	6

95. Samurai Sudoku X

5	8	1	3	2	7	6	4	9				3	4	2	5	9	8	6	7	1
7	4	3	9	6	5	8	2	1				1	9	8	6	7	4	3	5	2
6	2	9	4	1	8	7	3	5				5	6	7	2	3	1	4	8	9
2	5	8	1	4	3	9	6	7				4	2	3	8	6	9	7	1	5
3	9	4	7	8	6	1	5	2				7	1	5	4	2	3	8	9	6
1	6	7	5	9	2	4	8	3				6	8	9	7	1	5	2	4	3
8	7	6	2	5	9	3	1	4	8	7	9	2	5	6	9	4	7	1	3	8
9	1	5	6	3	4	2	7	8	1	6	5	9	3	4	1	8	2	5	6	7
4	3	2	8	7	1	5	9	6	4	3	2	8	7	1	3	5	6	9	2	4
						9	6	1	2	8	7	3	4	5						
						8	4	5	6	1	3	7	9	2						
						7	3	2	5	9	4	1	6	8						
5	6	7	3	2	4	1	8	9	7	4	6	5	2	3	1	9	7	4	6	8
4	9	1	8	7	5	6	2	3	9	5	1	4	8	7	5	2	6	1	3	9
3	2	8	9	1	6	4	5	7	3	2	8	6	1	9	4	3	8	5	7	2
2	1	9	4	5	8	3	7	6				7	6	8	3	5	9	2	1	4
8	5	6	7	3	9	2	1	4				9	4	5	7	1	2	6	8	3
7	3	4	1	6	2	5	9	8				1	3	2	6	8	4	9	5	7
1	8	5	6	9	3	7	4	2				8	5	4	2	6	3	7	9	1
9	7	3	2	4	1	8	6	5				3	7	6	9	4	1	8	2	5
6	4	2	5	8	7	9	3	1				2	9	1	8	7	5	3	4	6

96. Tetrominoku

1	1	4	3	3	2	4	2
4	2	2	3	1	4	1	3
3	3	1	2	4	4	2	1
2	4	3	2	1	1	3	4
1	2	1	4	4	3	3	2
3	4	3	4	2	2	1	1
2	3	4	1	3	1	2	4
4	1	2	1	2	3	4	3

97. Pentominoku

2	3	5	4	1	5	3	1	2	4
1	4	1	3	2	5	2	4	5	3
4	5	3	2	1	4	3	2	1	5
2	5	4	1	5	4	1	2	3	3
5	2	3	2	4	3	5	1	4	1
3	1	1	5	3	2	5	4	4	2
4	3	2	1	4	3	2	5	1	5
5	1	4	3	5	1	4	3	2	2
3	2	5	4	2	1	1	3	5	4
1	4	2	5	3	2	4	5	3	1

BIBLIOGRAPHY

1. D. Adams, *The Hitchhiker's Guide to the Galaxy,* Del Rey, New York, 1995; originally published 1979.
2. W. W. Adams, P. Loustaunau, *An Introduction to Gröbner Bases,* American Mathematical Society, Providence, 1994.
3. A. Adler, I. Adler, "Fundamental Transformations of Sudoku Grids," *Mathematical Spectrum,* Vol. 41, No. 1, 2008/2009, pp. 2–7.
4. K. Appel, W. Haken, "Every Planar Map is Four-Colorable," *Illinois Journal of Mathematics,* Vol. 21, 1977, pp. 439–567.
5. E. Arnold, S. Lucas, personal correspondence.
6. E. Arnold, S. Lucas, L. Taalman, "Gröbner Basis Representations of Sudoku," *College Mathematics Journal,* Vol. 41, No. 2, March 2010, pp. 101–112.
7. R. A. Bailey, Peter Cameron, Robert Connelly, "Sudoku, Gerechte Designs, Resolutions, Affine Space, Spreads, Reguli, and Hamming Codes," *American Mathematical Monthly,* Vol. 115, No. 8, 2008, pp. 383–404.
8. A. Baker, "Is There a Problem of Induction for Mathematics?" in *Mathematical Knowledge,* edited by M. Lang, A. Paseau, and M. Potter, Oxford University Press, New York, 2007, pp. 59–73.
9. A. Bartlett, T. Chartier, A. Langville, T. Rankin, "An Integer Programming Model for the Sudoku Problem," *The Journal of Online Mathematics,* Vol. 8, 2008, Article ID 1798, available online at http://www.maa.org/joma/Volume8/Bartlett/index.html, last accessed July 19, 2011.
10. Norman Biggs, E. Keith Lloyd, Robin Wilson, *Graph Theory, 1736–1936,* Oxford University Press, New York, 1986.
11. Robin Blankenship, Maxine Music, "Book Embeddings of Sudoku Graphs," Unpublished manuscript.
12. Kenneth Bogart, *Introductory Combinatorics,* 2nd Ed., Harcourt, Brace, Jovanovich, Publishers, San Diego, 1990, pp. 310–382.
13. J. Borwein, K. Devlin, *The Computer as Crucible: An Introduction to Experimental Mathematics,* A. K. Peters, New York, 2008.
14. R. C. Bose, S. S. Shrikande, E. T. Parker, "Further Results on the Construction of Mutually Orthogonal Latin Squares and the Falsity of Euler's Conjecture," *Canadian Journal of Mathematics,* Vol. 12, 1960, pp. 189–203.
15. Carl Boyer, *The History of the Calculus and its Conceptual Development,* Dover Publications, New York, 1949.
16. Marcel Danesi, *The Puzzle Instinct: The Meaning of Puzzles in Human Life,* Indiana University Press, Bloomington, 2002.

17. Jean-Paul Delahaye, "The Science Behind Sudoku," *Scientific American*, June 2006, pp. 80–87.
18. K. Devlin, "What is Experimental Mathematics," posted at http://www.maa.org/devlin/devlin_03_09.html, March 2009, last accessed on July 19, 2011.
19. K. Devlin, "When the Evidence Deceives Us," posted at http://www.maa.org/devlin/devlin_02_09.html, February 2009, last accessed on July 19, 2011.
20. Steven Dougherty, "A Coding-Theoretic Solution to the 36 Officer Problem," *Designs, Codes and Cryptography*, Vol. 4, No. 2, 1994, pp. 123–128.
21. William Dunham, *Journey Through Genius: The Great Theorems of Mathematics*, Penguin Books, New York, 1990.
22. Leonhard Euler, "Recherches sur une Nouvelle Espece de Quarres Magiques," *Opera Omnia*, Ser I, Vol. 7, 1782, pp. 291–392. English translation available at http://www.eulerarchive.org, last accessed on July 19, 2011.
23. Bertram Felgenhauer, Frazier Jarvis, "Mathematics of Sudoku, I," *Mathematical Spectrum*, Vol. 39, No. 1, 2006–2007, pp. 15–22.
24. A. Herzberg, M. Ram Murty, "Sudoku Squares and Chromatic Polynomials," *Notices of the American Mathematical Society*, Vol. 54, No. 6, June/July 2007, pp. 708–717.
25. Sophie Huczynska, "Powerline Communication and the 36 Officers Problem," *Philosophical Transactions of the Royal Society, A*, Vol. 364, 2006, pp. 3199–3214.
26. J. Gago-Vargas, I. Hartillo-Hermoso, J. Martín-Morales, J. M. Ucha-Enríquez, "Sudokus and Gröbner Bases: Noy Only a *Divertimento*," *Computer Algebra in Scientific Computing*, pp. 155–165, *Lecture Notes in Computer Science*, Vol. 4194, Springer, Berlin, 2006.
27. Dominic Klyve, Lee Stemkoski, "Graeco-Latin Squares and a Mistaken Conjecture of Euler," *The College Mathematics Journal*, Vol. 37, No. 1, Jan. 2006, pp. 2–15.
28. H. F. MacNeish, "Euler Squares," *Annals of Mathematics*, Vol. 23, 1922, pp. 221–227.
29. G. McGuire, "Sudoku Checker and the Minimum Number of Clues Problem," posted at http://www.math.ie/checker.html, November 2006, last accessed July 19, 2011.
30. Michael Mepham, "Solving Sudoku," posted at http://www.sudoku.org.uk/PDF/Solving_Sudoku.pdf, last accessed July 19, 2011.
31. Bastian Michel, "Mathematics of NRC-Sudoku," preprint posted at http://www.staff.science.uu.nl/~kalle101/webfiles/sudoku.pdf, last accessed July 19, 2011.
32. P. Riley, L. Taalman, *Naked Sudoku*, Puzzle Wright Press, New York, 2009.
33. P. Riley, L. Taalman, *No-Frills Sudoku*, Puzzle Wright Press, New York, 2011.
34. Jason Rosenhouse, *The Monty Hall Problem: The Remarkable Story of Math's Most Contentious Brain Teaser*, Oxford University Press, New York, 2009.
35. G. Royle, "Minimum Sudoku," posted at http://mapleta.maths.uwa.edu.au/~gordon/sudokumin.php, last accessed July 19, 2011.
36. E. Russell, F. Jarvis, "Mathematics of Sudoku II," *Mathematical Spectrum*, Vol. 39, No. 2, 2006/2007, pp. 54–58.
37. R. van der Werf, "Minimum Sudoku X Collection," posted at http://www.sudocue.net/minx.php, last accessed July 19, 2011.
38. Ian Stewart, "Euler's Revolution," *New Scientist*, March 24, 2007, pp. 48–51.
39. Douglas Stinson, *Combinatorial Designs*, Springer, New York, 2004.
40. Douglas Stinson, "A Short Proof of the Nonexistence of a Pair of Orthogonal Latin Squares of Order Six," *Journal of Combinatorial Theory, Series A*, Vol. 36, 1984, pp. 373–376.

41. Sudoku Programmers Forum, posted at http://www.setbb.com/phpbb/?mforum=sudoku, last accessed July 19, 2011.
42. J. J. Sylvester, "Chemistry and Algebra," *Nature*, Vol. 17, p. 284, 1878.
43. L. Taalman, "Taking Sudoku Seriously," *Math Horizons*, September 2007, pp. 5–9.
44. Gaston Tarry, "Le Probleme de 36 Officiers," *Compte Rendu de l'Assoc. Francais Avanc. Sci. Naturel*, Vol. 2, 1900, Vol. 1, 1900, pp. 122–123.
45. Gaston Tarry, "Le Probleme de 36 Officiers," *Compte Rendu de l'Assoc. Francais Avanc. Sci. Naturel*, Vol. 2, 1901, pp. 170–203.
46. Robin Wilson, *Four Colors Suffice: How the Map Problem Was Solved*, Princeton University Press, Princeton, 2002.
47. D. Wolpert, W. Macready, "The No Free Lunch Theorems for Optimization," *IEEE Transactions on Evolutionary Computation*, Vol. 1, No. 1, pp. 67–82, 1997.

INDEX